T0281532

Physics of Data Science
and Machine Learning

Physics of Data Science and Machine Learning

Ijaz A. Rauf
Adjunct Professor
School of Graduate Studies, York University, Toronto, Canada
Associate Researcher
Ryerson University, Toronto, Canada
President
Eminent-Tech Corporation, Bradford, ON, Canada

CRC Press
Taylor & Francis Group
Boca Raton London New York

CRC Press is an imprint of the
Taylor & Francis Group, an **informa** business

First edition published 2022
by CRC Press
6000 Broken Sound Parkway NW, Suite 300, Boca Raton, FL 33487-2742

and by CRC Press
2 Park Square, Milton Park, Abingdon, Oxon, OX14 4RN

© 2022 Taylor & Francis Group, LLC

CRC Press is an imprint of Taylor & Francis Group, LLC

Library of Congress Cataloging-in-Publication Data
Names: Rauf, Ijaz A., author.
Title: Physics of data science and machine learning / Ijaz A. Rauf.
Description: First edition. | Boca Raton : CRC Press, 2022. | Includes
bibliographical references and index.
Identifiers: LCCN 2021023415 | ISBN 9780367768584 (hardback) |
ISBN 9781032074016 (paperback) | ISBN 9781003206743 (ebook)
Subjects: LCSH: Physics—Data processing. | Physics—Methodology. | Machine
learning. | Data mining. | Statistical mechanics. | Quantum statistics.|
Probabilities. | Mathematical optimization.
Classification: LCC QC52 .R38 2022 | DDC 530.0285/53—dc23
LC record available at https://lccn.loc.gov/2021023415

ISBN: 978-0-367-76858-4 (hbk)
ISBN: 978-1-032-07401-6 (pbk)
ISBN: 978-1-003-20674-3 (ebk)

DOI: 10.1201/9781003206743

Typeset in Minion
by codeMantra

Dedicated to:

My beautiful wife, Najiba, and our four wonderful children, Shahrukh, Sherjeel, Shahroze, and Rumessa, everyone admirable, smart, and outstanding in their unique ways, and to Rosemah, who is now part of our family by marrying Shahrukh.
And
to all those brave physicists who like to challenge the conventional way of doing physics.

Contents

Preface: Motivation and Rationale

Physicists first realized as early as the seventeenth century that material systems can be described by a small number of descriptive parameters related to one another through formulae, explaining natural laws. This use of descriptive parameters gave birth to what later became known as statistical physics that referred to parameters such as geometric, dynamic, and thermodynamic properties of materials.

The older theory had no probabilistic qualifications to its laws. Later, Maxwell and Boltzmann introduced the concept of equilibrium in exciting ways that generated a new "statistical thermodynamics," and imported the new probabilistic structures into the older theory. In the 1900s, Max Plank brought in additional concepts to the mix, introducing a discrete way of explaining physical phenomena, changing to a quantum mechanical theory.

Though physicists' fundamental contributions to the development of basic principles and theories in both statistical and quantum mechanics, diversification, and specialization started to give birth to new fields and areas of focus in science, there have been significant advancements by mathematicians, statisticians, and computer scientists in developing applied methodologies that enhance the speed of knowledge discovery and scientific advancement. Although one can find some physics courses on Bayesian analysis, methodologies such as neural networks, and the design of experiments, though readily adopted by biologists, chemists, and engineers, are only mentioned in research papers for physicists.

Physicists are the most suitable candidates, as they excel in the emerging fields of data mining, data science, machine learning, and artificial intelligence through self-learning. The advanced concepts of quantum and statistical mechanics that form the basis of statistical design of experiments and neural networks seldom find their way into any physics degree requirements, even at institutes that offer an engineering physics degree.

It is hard to find any book written explicitly for physicists linking the fundamental concepts of quantum and statistical mechanics to modern data mining, data science, and machine learning. This book is expected to be the first of its kind written explicitly for physicists explaining the concepts of design of experiments, neural networks, and machine learning, building on fundamental concepts of statistical and quantum mechanics.

Even though I graduated from Cambridge University with a Ph.D. in physics in 1991, I had no clue about the Statistical Design of Experiments. My Ph.D. was in experimental physics, and all the experiments were performed by varying one factor at a time while keeping all the others constant. This one factor at a time remained the way of experimenting for many of my colleagues and me when I worked in physics departments at Queen's University, the University of Alberta, and the University of Illinois at Chicago (UIC). In the year 2000, I left the academic world. I joined PerkinElmer Optoelectronics as a senior scientist. I was introduced to the Statistical Design of Experiments concepts when I received my Master Black Belt Certification in Six Sigma Methodologies. Although initially skeptical, as a hardcore physicist, but seeing the results during application, I fell in love with the Design of Experiments and later the machine learning concepts because of the speed of experimentation and knowledge development it provided. I have always tried to introduce these concepts to physicists by offering introductory courses at York University, where I am an adjunct professor. However, these courses have usually been more appealing to engineers than physicists; still few pioneering physicists have shown interest in these concepts, encouraging the writing of this book.

This book is expected to generate a new trend among research physicists to utilize these techniques to advance and discover new knowledge. This book also provides a self-learning tool for physicists interested in a career in data science, machine learning, and artificial intelligence.

Ijaz A. Rauf
March 2021

Author

 **Dr. Ijaz A. Rauf** is a physicist, data scientist, and a Lean Six Sigma Master Black Belt (MBB). He is an innovator, experienced consultant, educator, and a seasoned program manager. Dr. Rauf obtained a Ph.D. in physics from Cambridge University, England. He started his career as a consultant for a small business, while working as a researcher and then educator at various universities in Canada and the United States. After spending eight years in academia, he moved to the high-tech industry before starting his own businesses.

Dr. Rauf has founded and cofounded three high-tech companies, has served as the President and CEO at SolarGrid Energy Inc., and is currently managing his own data science and Lean Six Sigma consulting company, Eminent-Tech Corporation. Besides leading his business ventures, he is also an adjunct professor in the School of Graduate Studies, Department of Physics and Astronomy, York University, and a research associate at Ryerson University in Toronto, Canada.

Dr. Rauf has been awarded patents for industrial designs and processes invented during his career, and has authored more than 50 international scientific publications. He has served as a council member for the Ontario College of Teachers, vice-chair for Blue Hills Child and Family Centre's board of directors, and a panel expert for the International Renewable Energy Agency (IRENA). He has also served as a visiting professor at the University of Energy and Natural Resources in Ghana and as foreign faculty at COMSATS Institute for Information Technology in Pakistan. He is also a member of the International Scientific Advisory Board for the University of Energy and Natural Resources, Ghana.

Dr. Rauf's own research focuses on predictive analytics and machine learning. Besides data science and Lean Six Sigma, Dr. Rauf also has expertise in the areas of nanotechnology, thin films, and nanomaterials, especially for their applications in solar cells and nanobiotechnology-based sensors and detectors for point-of-care diagnostics.

CHAPTER 1

Introduction

1.1 A PHYSICIST'S VIEW OF THE NATURAL WORLD AND PROBABILITIES

Physicists aim to find a single theory that can explain the whole universe, but to do so, they must solve some of the most challenging problems of science. Before Newton, a physicist's view of the world was only two-dimensional. By introducing gravity, Newton enabled us to pass from a picture with two-dimensional symmetry to a picture with three-dimensional symmetry. Building on this, Einstein expanded this three-dimensional view of the world to a four-dimensional view by introducing the time domain. The time domain was introducing the particular theory of relativity that considers the four-dimensional world view to be linear. Einstein further expanded this view by proposing relativity's general theory, where the space–time continuum is curved (or nonlinear). The general theory of relativity helps explain most of the physical phenomenon; however, it fails when phenomenon at microscopic scales (atomic scales) is considered. The atomic-scale phenomenon is the area that leads to the emergence of the quantum theory, and it pertains to the discussion of tiny things – subatomic phenomenon.

The quantum theory first emerged when Planck realized the need to conclude that electromagnetic waves' energy can only occur in multiples of a certain amount; this depended on the waves' frequency to justify the law of black-body radiation. Then, Einstein discovered that the same energy unit occurs in the photoelectric effect. Bohr's atomic model provided an incomplete and very primitive picture of the physical world on a small scale. The quantum theory's significant breakthrough came in 1925

DOI: 10.1201/9781003206743-1

when quantum mechanics was advanced by Heisenberg and Schrodinger independently. This development resulted in a dramatic shift in the image of the universe by the physicist because we had to give up the deterministic image that we often took for granted. We are guided to a hypothesis that does not predict with certainty what is expected to happen in the future but instead gives us knowledge about the likelihood (probability) of occurrence of different events.

Physicists are used to working with massive amounts of data, but they are lucky because their experimental data are of very high quality. The advent of big data has resulted in an influx of data sets with a much more complicated structure – a trend needing new methods and a different mindset. The growth of big data offers an opportunity for physicists. Nevertheless, they need a slight yet necessary change in attitude to take full advantage. Physicists like to think they understand the data to figure out the causal relationships between events; they have scientific methods, mathematical tools, and analytical knowledge. Some data structures are more complex than others. However, if correctly performed, well-designed experiments produce clean data that can be analyzed to draw simple, empirical conclusions about nature's inner workings. However, this is exceptionally wishful thinking as it tends to downplay the role of experimental design.

The experimental design is where physicists excel by working in a natural system where clean experiments are possible. A strong understanding of calculus, arithmetic, and programming helps. Physicists are comparatively superior from a technological viewpoint compared with their colleagues in other natural sciences. The truth is that statisticians, computer scientists, biologists, psychologists, and economists frequently lack the privilege of working with clean data.

An increasing number of problems come in a shape opposite of what physicists are used to seeing, for example, to evaluate user experiences on individual websites or social media or assess epidemics' dynamics to determine the consequences of economic policy, which can be quantified meaningfully. These are situations where "experiments" are far from optimal but relatively easy to conduct and generate data streams with an incredibly complex structure.

Physicists must learn a new lexicon to translate their expertise to be useful in new unknown circumstances. An essential move in this process is to acknowledge that while models have a high degree of applicability to evaluate large-scale data sets, their use is just one approach. Physicists tend to opt for generative models, namely those that allow synthetic data generation before any observation. Nevertheless, a depiction of nature relies on numerous assumptions and approximations. In comparison, discriminative models do not have a mechanism for generating the data. These are based on a series of techniques that view the experimental data as a direct input, which is then used to refine the fitting model iteratively. This method is possible thanks to Bayes' theorem.

There is also plenty that physicists can learn from modern statistics and probability, and the information theory. Complemented by a few primary principles from these disciplines, the techniques already mastered by physicists from statistical mechanics and field theory can be adapted for use in the most complex data analysis tasks facing other disciplines of science. It may be surprising that mathematical techniques initially developed in physics, including those necessary to calculate the partition function, were transferred and further developed into other research domains. It is time that physicists to learn and reclaim these advancements.

1.2 DATA – TYPES OF DATA

Humans have collected, processed, and analyzed data from time immemorial without developing a name for this. It was 1640 when the word "data" first appeared in the English language when physicists realized that the material systems are better understood when described in a small number of parameters. Values prescribed for these parameters became to be known as data. Data can broadly be divided into two main classes: quantitative and qualitative data. Both can be further divided into two subclasses, as shown in Figure 1.1. The quantitative data could be continuous,

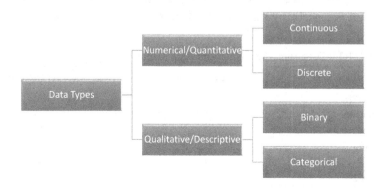

FIGURE 1.1   Various classifications of data.

where it could be measured on an infinitely divisible scale provided we have sensitive enough measuring equipment. For example, the distance between two points, the length, and height of an object are all examples of continuous data. Alternatively, it could be discrete when it can only have specific values, such as the number of students in the class and the number of electrons in an atom.

On the other hand, the qualitative data can be either binary. It can have only one of the two prescribed values: the charge on a particle (positive or negative) and exam result (pass or fail). Alternatively, it could be an attribute, for example, color, religion, or nationality.

1.2.1 Data to Information

One may drive some meanings from given data itself when there is only a small amount of data. As the amount of data starts to increase, it becomes increasingly difficult to derive meaning from the data. Table 1.1 shows data from a semiconductor manufacturing line showing various causes for chip failure. Data in its raw form require a bit of human effort to sort through and get the information. However, if these data are transformed into Pareto charts, as shown in Figure 1.2, significant information becomes readily available for decision-making by the observer.

The term "information" is defined as the data presented in context, structured, organized, and processed, making it meaningful and useful to the individual who needs it. Data mean raw facts and statistics about people, locations, or something else represented as numbers, letters, or symbols.

Information is the converted and classified data into an intelligible form that can be used in the decision-making process. In short, when the data after processing turn out to be significant, it is called knowledge. In essence, it is something that tells; it addresses a particular query.

TABLE 1.1   Data from a Semiconductor Chip Manufacturing Line

Defect Type	Frequency (Count)	Loss Per Defect ($)
Cracking	12	100
Soldering	33	300
Others	3	50
Etching	27	600
Pinhole	4	200
Flacking	8	150
Molding	13	250

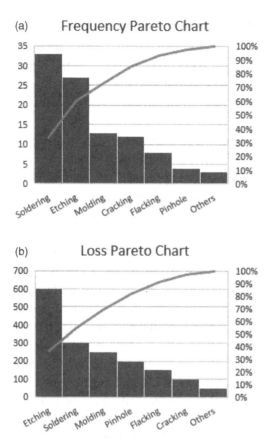

(a) **Frequency Pareto Chart**

(b) **Loss Pareto Chart**

FIGURE 1.2   Data from Table 1.1 transformed into information for the observer for better decision-making.

Accuracy, relevance, completeness, and availability are the key features of the knowledge. It can be conveyed in the form of a message text or by observation. It can be obtained from various sources such as newspapers, television, the internet, people, and books.

As the amount of data increases, the transformation required to derive information from the data for meaningful decision-making becomes increasingly complex and challenging, requiring advanced statistical and graphical tools. A large amount of information can be transformed into knowledge by understanding the underlying relationships that help generalize the formulae or the theoretical framework for the phenomenon that generated the data. Converting information to useful knowledge is even more difficult and complicated, requiring tools to develop models for the systems under observation.

1.2.2 Information to Knowledge

Knowledge means a person's insight and understanding, location, activities, thoughts, queries, ways of doing things, or something learned through studying, perceiving, or discovering. This new learning is the state of cognizance, recognizing something through the comprehension of concepts, research, and experience.

In short, information connotes an entity's firm theoretical or realistic understanding, along with its ability to use it for a specific reason. A combination of information, experience, and intuition leads to knowledge that can draw inferences and develop insights based on our expertise. This knowledge also helps in decision-making and action-taking.

1.2.3 Critical Differences between Information and Knowledge

The points given below are essential. The difference between information and knowledge is concerned: The information denotes the organized data about someone or something obtained from various sources such as newspapers, the internet, television, and discussions. Knowledge refers to the understanding or comprehension of the acquired subject from a person's education or experience. Knowledge is nothing but the distilled data type that helps to understand the context. Knowledge, on the other hand, is the appropriate and objective information that helps to conclude.

Information is generated by data gathered within the appropriate context. Conversely, it results in knowledge when the information is paired with experience and intuition. Processing enhances the representation, thus ensuring a simple understanding of the information; despite this, processing results in enhanced knowledge, thereby increasing the subject's understanding. The information helps comprehend the facts and figures, unlike the knowledge that contributes to the subject's understanding. The transmission of information is secure by different means, that is, verbal or nonverbal signals. Conversely, knowledge transfer is a little complicated, as it requires the user to develop an understanding.

Information can be reproduced at a low cost. However, precisely similar knowledge reproduction is not possible because it is based on experiential or individual values and perceptions. It is not enough to use information alone to make generalizations or predictions about someone or something. Knowledge may, on the contrary, predict or draw inferences. Every piece of information is not necessarily a piece of knowledge, but all knowledge is based on information.

1.3 DATA MINING FOR KNOWLEDGE

As the name goes, the process of data mining could be understood by comparing it to traditional mining operations, where we dig through vast amounts of dirt and stones to find the gems hidden deep down. Data mining is thus the process of digging through a sea of raw data to discover new and previously unknown insights and information. Data mining is part of a more comprehensive knowledge discovery process, including data extraction, data cleaning, data fusion, data reduction, and feature creation, known as preprocessing and postprocessing steps such as pattern and model analysis, confirmation, and generation of hypotheses. The information creation and the method of data mining appear to be strongly incremental and interactive.

Data mining is a process based on using tools to extract useful information that can be transformed into knowledge from large data sets and extend human knowledge. Thus, data mining tools could help a person or an organization discover hidden knowledge in enormous data. Data mining refers to discovering new patterns from a wealth of data in databases by focusing on algorithms to extract useful knowledge. Data mining is the mechanism of discovering insightful, exciting, and novel patterns, as well as descriptive, understandable, and predictive models for the process of interest from large-scale data.

1.4 MACHINE LEARNING AND ARTIFICIAL INTELLIGENCE

Learning, in general, is the process of gaining knowledge or skills through study, education and training, observation, or experimentation. Learning is a procedure that leads to a permanent capacity and or capability change in living beings. Learning through personal experience and knowledge is at the root of human intelligence, propagating from generation to generation. Learning shapes our brains and develops our intellect that translates into intelligence. Social learning comes from experience, observations, interactions, and traditional education and training. In an institutionalized setup, human beings are taught the concepts; then, examples help them solve the concepts. The human capacity or skills developed through this exercise of knowledge transmission are then evaluated by providing them related but a new set of problems that use the same concepts to solve. In this way of evaluation, the learner's ability to generalize is tested.

A problem may also be solved by a machine that learns. Generalization is perhaps the best way machine learning can be conceptualized. Machine learning is emerging as an umbrella noun for methods studied and developed for many decades in different scientific communities, such as statistical learning, pattern recognition, image processing and analysis, computer vision, and computational learning, among many others. A machine processes the data and automatically finds structures in the data, i.e. learns. The knowledge about the extracted structure in the data is then used to solve the problems at hand. This approach of problem-solving is called inductive. Machine learning is about inductively solving problems by machines, i.e. computers. The ability to provide accurate predictions and forecasts stems from machine learning models by considering real-time input data and historical data. Although an accurate prediction or forecast is invaluable, it is priceless to make analytics-driven decisions about the best course of action to be taken.

Artificial intelligence (AI) is referred to as machines' ability to mimic the human trait of decision-making based on data analyses and prior knowledge. AI took birth with the invention of the mechanical calculators in the mid-seventeenth century and has advanced significantly to the invention of the first mechanical computer (historically known as Jacquard Loom) in 1804–1805. AI took a new turn with the advent of solid-state electronic calculators in the 1960s and then modern computers in the early 1970s. The definition of AI continues to evolve, as many operations (for example, simple calculations as performed by calculators or computers) are now taken for granted.

The modern definition of AI assumes that human intelligence can be defined so that machines can simulate it. Machine simulation of human intelligence can be accomplished by feeding the learning and the prediction or forecasts generated by machine learning models as inputs to a decision optimization model. The decision optimization model can then consider the various trade-offs and constraints to develop the optimal solution for the situation under consideration and automatically trigger an allowed action. Subsequently, once an optimization model has triggered or recommended an action plan and that plan is in operation, the data from the execution operations of that plan can be used by machine learning models. These new data help the machines improve predictions and forecasts to make the decision models more precise and accurate automatically.

The ability of AI to make decisions, refine the models through constant learning to improve the precision and accuracy, minimize the risks involved faster than human beings, plus the possibility of eliminating human errors, makes it very attractive for a myriad of applications, from simple applications such as spell checkers, autocorrections, Alexa, Siri, and video games to more complex applications such as autonomous vehicles for human transportation, AI-assisted medical diagnostics, and AI-based law clerks, and the list continues. As technology advances further, even more applications come to the surface that can use AI for much faster, precise, and accurate actions.

1.5 SCOPE: WHAT THIS TEXT COVERS

This book is written for postgraduate levels, and thus to understand this book, a learner needs to understand sciences at the undergraduate level, especially in fundamentals of physics and statistics.

For a physicist, this text is designed to review fundamental physics concepts and build on these concepts through the fundamentals of probability theory leading to the modern concepts in data science and machine learning, opening the way for a deeper understanding of both sides of the equation.

For statisticians and data scientists, it provides a basic understanding of the underlying physics that has led to the modern tools, which are so popular with businesses. They can understand these fundamental concepts of physics and build on these to solve problems in the areas where modern data science, machine learning, AI, and the internet of things are facing challenges.

This book introduces many concepts for physicists, statisticians, and data scientists that may incite further studies into more complex and advanced topics in these areas.

Following this brief introductory chapter, Chapter 2 introduces Newtonian mechanics starting with fundamental laws of motion, concept of center of mass, and the law of universal gravitation. Then, Lagrangian and Hamiltonian mechanics are discussed before classical field theory is described, and based on all these concepts, Maxwell–Boltzmann equilibrium is introduced. Chapter 3 focuses on the quantum mechanics where concepts such as kinematical framework, wave functions, and Heisenberg's uncertainty principle are introduced. Discussion on boundary conditions, energy states, and quantum confinement is followed by introduction of harmonic oscillator, and probability density including the concepts of expected value and finally quantum field theory is described.

Chapter 4 is focused on probabilistic physics, where concepts such as probability theory that included expected value, probability amplitude and quantum interference are introduced. Probability distributions are described before introducing the central limit theorem that is followed by description of hypothesis testing, confidence level, and confidence interval.

Chapter 5 carries detailed practical discussions on the statistical design of experiments. Starting with measurement systems analysis, concepts of random and systematic errors are introduced. Regression analysis and analysis of variance are discussed, which lead to the concept of statistical design of experiments and systems modeling.

Starting with the basics of information theory, Chapter 6 introduces machine learning including the concepts of supervised and unsupervised learning, modeling input and output functions, and Bayesian decision theory. Finally, neural networks and application to classification problem are discussed. Modern machine learning techniques make it possible to develop models that accurately depict the physical systems in a digital world, known as digital twins, a concept that is discussed at the start of Chapter 7. Later parts of this chapter discuss the simulation and optimization techniques developed to take advantage of digital models. Concepts for verification and validation of the models along with automation of verification and validation process are discussed at the end of this chapter.

1.5.1 What This Text Does Not Cover

This text is not a specific course in itself, or a review of such a course. It does not cover all basic concepts in fundamentals of physics or data science and machine learning; it covers only those that are very important or especially useful for both sides of the equation.

The detailed derivation of physics' fundamental concepts is not provided as that was thought to be out of scope for this book, as many texts already exist where such derivations can be found. However, some basic concepts in probability theory that lead to the concepts of design of experiments are derived for physicists.

CHAPTER 2

An Overview of Classical Mechanics

THE PHYSICAL SCIENCE THAT studies the displacement of bodies under the motion of forces is classical mechanics. The modern age of mechanics was introduced by Galileo Galilee, using mathematics to explain the movement of bodies. He added the idea of force and defined the continuously accelerated motion of objects near the Earth's surface. Seventy years later, Newton developed his Laws of Motion and addressed the most significant scientific dilemma of his time by applying his Universal Law of Gravitation to establish planetary motion.

Because Newton's laws of motion provide the basis for classical mechanics, it is also called Newtonian mechanics. Mechanics has two parts: kinematics and dynamics. Kinematics deals with the geometric definition of object motion without taking into account the forces causing the movement. Dynamics is the component concerned with the details causing changes in motion or changes in other properties. This component leads us to the concept of force, mass, and the laws that regulate the movement of objects.

The equations developed before 1900 were still ideally suited to describe objects of typical sizes and velocities. The concepts of energy gradually evolved, resulting in the discovery of energy conservation theory and its subsequent application to the laws of thermodynamics. Conservation theories are now fundamental to our understanding of mechanics; conserving momentum, energy, and angular momentum have allowed Newtonian mechanics to be reformulated. The reformulation occurred to apply the

DOI: 10.1201/9781003206743-2

laws to different situations, such as the Lagrange, the Hamilton, and the Hamilton–Jacobi theoretical systems.

Italian French mathematician and astronomer Joseph-Louis Lagrange developed a mathematical technique known as the Lagrangian that is a function of the generalized coordinates, their time derivatives, and time and contains details about system dynamics. Lagrangian mechanics is suitable for conservative force systems and, in any coordinate system, for bypassing constraint forces. It is possible to compensate for dissipative and induced effects by splitting the external forces into a total of potential and nonpotential strengths, resulting in a set of modified Euler–Lagrange (EL) equations. For simplicity, EL equations may use generalized coordinates to leverage symmetries in the system or the geometry of the constraints that can simplify the system's motion solution. As a case of Noether's theorem, Lagrangian mechanics also shows conserved quantities and their symmetries directly. Lagrangian mechanics is essential for its wide-ranging applications and for its role in advancing profound physics understanding.

The Hamilton mechanics is the formulation of the theory of static action by William Rowan Hamilton. He claims that a physical system's dynamics are defined by a fundamental equation based on a single function, which incorporates all the physical details about the system and the forces acting on it. The variational problem is analogous to and permits the derivation of the physical system's differential equations of motion. While initially formulated for classical mechanics, Hamilton's principle still applies to physical fields such as the electromagnetic and gravitational fields. It plays a vital role in the theories of quantum mechanics, quantum field theory, and criticality.

2.1 NEWTONIAN MECHANICS

In Newtonian mechanics, a given body's movement is evaluated in terms of the forces acting upon it. Space and time are the two fundamental principles for the study of motion, all of which are continuous variables. A body's location in space must be defined as a function of time to understand and describe its motion. A coordinate system is used as a reference frame for this reason. One simple method that we often use is the Cartesian coordinate method. The position of a point P in a Cartesian coordinate system is specified by three coordinates (x, y, z) or spherical polar coordinates (r, θ, ϕ), as shown in Figure 2.1. The particle P motion can then be studied by the change in the position coordinates over time.

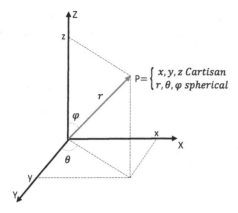

FIGURE 2.1 Cartesian and spherical coordinate system to describe the position of a particle.

Kinematics is primarily concerned with motion mechanics, but dynamics attempts to address the question as to what movement can occur when defined forces act upon a body. Newton's laws of motion are the rules that allow one to make the relation. They are physics laws that are founded on scientific data and stand or fall because their predictions are correct.

2.1.1 Newton's Laws of Motion

First Law of Motion: An object remains at rest unless acted up by an unbalanced force. Newton's first law suggests that no external force is required to sustain a body's uniform motion; it continues unchanged because of the body's intrinsic property, which we call inertia. Because of this, the first law of motion is also called the law of inertia. Inertia is a body's normal inclination to remain at rest or in continuous motion in a straight line. Quantitatively, a body's inertia is determined by its mass (m) and its velocity (v), that is, its momentum. A body's momentum (p) is proportional to its velocity. In mathematical terms, Newton's first law can be expressed as follows: In the absence of an unbalanced external force, $p = mv =$ constant; this is the law of momentum conservation.

Second Law of Motion: A net force acting on an object can change the momentum of the object, the rate of which is proportional to and in the direction of the net force.

$$F = \frac{dp}{dt} = m\frac{dv}{dt} = m\frac{d^2r}{dt^2} = ma \tag{2.1}$$

where *a* stands for acceleration and *r* is the position vector in the Cartesian space. Equation 2.1 is also called the particle's motion equation. It is a differential equation of the second order. Provided force *F*, the position and velocity of a particle are known at a specific time; the particle's position and velocity can be found at any other time with the help of the second law. It means that the correct values of coordinates and velocity (or momentum $p = mv$) must be known simultaneously at a given moment to understand the particle's trajectory fully.

Third Law of Motion: When an object exerts a force on another object, there is an equal but opposite in direction force exerted simultaneously by the second object on the first. In two particles (1 and 2), the force F_{12} applied by particle 1 on particle 2 is equal but opposite in direction to the force F_{21} exerted by particle 2 on particle 1. In other words:

$$F_{12} = -F_{21} \ or \ \frac{dp_1}{dt} = -\frac{dp_2}{dt} \ or \ m_1 a_1 = -m_2 a_2 \tag{2.2}$$

2.1.2 Angular Momentum, Work, and Energy

Our discussion so far has been about particle systems with linear motion. In systems with rotation motion, torque and angular momentum are two equivalent essential parameters. A force induces linear acceleration, while a torque causes angular acceleration. A particle's angular momentum about a point O (say origin), denoted by *L*, is defined as $L = r \times p$, where *r* is particle's displacement vector from the origin "O." The expression for the torque Γ or the moment of the force about the origin is

$$\Gamma = r \times F = r \times \frac{dp}{dt} = \frac{d(r \times p)}{dt} = \frac{dL}{dt} \tag{2.3}$$

The law of conservation of angular momentum dictates that if the torque acting on an object in a circular motion is zero, then the angular momentum is constant.

$$\Gamma = 0 = \frac{dL}{dt} \text{ implies } L = \text{Constant} \tag{2.4}$$

The most common examples of constant angular momentum are planets traveling around the Sun and satellites circling the Earth.

Considering the work done by a force leads us to the derivation of the law of energy conservation. Work done by an external force in moving a particle from position 1 to position 2 is given by the force's scalar product and the displacement vectors.

$$W_{12} = \int_1^2 F.dr = \int_1^2 m\frac{dv}{dt}.dr = m\int_1^2 v\ dv = \frac{m}{2}\left(v_2^2 - v_1^2\right) = K_2 - K_1 \quad (2.5)$$

where K_2 and K_1 are the particle's kinetic energies at positions 1 and 2, and work is performed by force on the particle if $K_2 > K_1$. As a result, the kinetic energy of the particle increases. If $K_2 < K_1$, the work is done by the particle against the force; thus, the kinetic energy of the particle decreases. If the force acting on a particle is such that the work carried out along a closed path is zero, then the force is conservative. For a conservative force F, $\oint F.dr = 0$ based on stokes theorem, if the closed path encloses a surface "s" then $\oint F.dr = \int_s (\nabla \times F).ds = 0$, for an arbitrary surface, $\nabla \times F = CurlF = 0$, which is a necessary condition for a conservative force. The curl of a vector is zero if it can be expressed as a gradient of a scalar function of position; hence, $F = -\nabla U(r)$. This scalar function $U(r)$ is called the particle's potential energy at the position represented by vector r. If the force F is a conservative force, the work done can be described as:

$$W_{12} = \int_1^2 F.dr = -\int_1^2 U(r).dr = U_1 - U_2 \quad (2.6)$$

From Equations 2.5 and 2.6, we get $K_1 + U_1 = K_2 + U_2$, which describes the law of conservation of energy mathematically. The potential energy introduced here is not unique, and thus the absolute potential value has no significance. Likewise, the kinetic energy also has no absolute value since we use an inertial reference frame to calculate the velocity and, thus, the kinetic energy.

2.1.3 Multiple Interactions and Center of Mass

The dynamics of a system of particles can be studied using a simple application of Newton's laws. In addition to the externally applied forces, this application of Newton's laws considers the forces acting between the particles. One can easily use the Newtonian mechanics' principles of a single-particle to a multiparticle system, too. Let us consider a system of n particles P_1, P_2, ...P_n all interacting with each other. Supposing all other influences are removed from the system, then acceleration a_0 induced by the interaction from the rest of the particles is given as:

$$a_0 = a_{01} + a_{02} + ... + a_{0n} \qquad (2.7)$$

where a_{01} and a_{02} are accelerations induced on P_0 as if P_1 and P_2 were individually interacting with P_0. Multiplying both sides of Equation 2.7 with mass m_0 of the particle P_0 gives us:

$$m_0 a_0 = m_0 a_{01} + m_0 a_{02} + ... + m_0 a_{0n} \text{ or } F_0 = F_{01} + F_{02} + ... + F_{0n}$$

Therefore, Newton's second law remains valid for multiple interactions, provided the "resultant force" F_0 acting on P_0 is understood to mean the (vector) resulting from the individual forces acting on P_0. If the position vectors of this system of n particles are given as r_1, r_2, ... r_n, respectively, then the center of mass of this system of n particles is a point in space whose position vector R is given as:

$$R = \frac{m_1 r_1 + m_2 r_2 + ... + m_n r_n}{m_1 + m_2 + ... + m_n} = \frac{\sum_{i=1}^{n} m_i r_i}{\sum_{i=1}^{n} m_i} = \frac{\sum_{i=1}^{n} m_i r_i}{M} \qquad (2.8)$$

The center of mass for a particle system is essentially a "weighted" mean of the particle position vectors, where the particle masses are the "weights." Center of mass is an essential concept in multiparticle system mechanics.

2.1.4 The Law of Gravitation

Physicists only consider four distinct kinds of forces of interaction that exist in nature. These are gravity forces, electromagnetic forces, and weak/strong nuclear forces. The nuclear forces are only essential inside the atomic nucleus, and they will not affect us here. The most remarkable achievement of Newton was to eventually discover the law of universal gravity and solve the question of two centuries of planetary motion. Kepler started the conception of gravitation as an attractive force between two bodies. He also suggested the gravitation of the Sun would hold the planets orbiting around the Sun. However, it was mainly Hooke (and Halley and Wren) who successfully demonstrated that gravitational force was an attractive force toward the Sun. Newton took up the gravitation problem and proved that the second law of Kepler is valid for every central force, not just for the inverse square force due to gravitational attraction.

Newton's law of gravitation postulates that any object with mass attracts any other object with mass, with force called gravitation. The third law stipulates that when gravitational interaction occurs between particles, the interaction forces must be equal in magnitude, opposite in direction, and parallel to the straight line that separates the particles. Equation 2.9 sets the magnitude of the forces of gravitational interaction:

$$F_{12} = F_{21} = G\frac{m_1 m_2}{R^2} \qquad (2.9)$$

Here m_1 and m_2 are masses of the particles, R is the distance between them, and G is the universal gravitational constant. G is not dimensionless; its numerical value depends on the units of mass, length, and force. In SI units, the constant of gravitation is given by $G = 6.67 \times 10^{-11}$ Nm²/kg². The same is true for a mass distributed over a large volume; for example, if M is the mass of the Earth that is spread over the radius R of the Earth, it can be treated as a point mass, M being placed at the center of the planet. Thus a small particle of mass m experiences a force given as:

$$F = G\frac{Mm}{R^2} = mg \text{ or } g = G\frac{M}{R^2} \qquad (2.10)$$

Equation 2.14 shows that the acceleration g experienced by any object on the surface of the Earth is independent of its mass and is equal to 9.8 m/s².

2.2 LAGRANGIAN MECHANICS

During previous sections, we illustrated the usefulness of Newton's laws of motion in addressing several issues. Nevertheless, if the system is subject to external constraints, it can be challenging to solve the equations of motion, and often it can even be challenging to formulate them. The forces of constraints are typically very complex or unknown, which hinders formalism. Two separate approaches, Lagrange's and Hamilton's formulas, were created to solve these difficulties. Both methods are constructed based on an energy approach so that Newtonian formalism derives from it.

2.2.1 Constraints

A constrained object is not able to move in any form freely. The conditions that restrict the object's motion are called constraints. For example, in a container, the gas molecules are limited by the vessel's walls to travel only within the box. Mainly there are two types of constraints: holonomic and nonholonomic. When a system has constraints that are only geometric or integrable kinematic, it is holonomic. When it does have nonintegrable cinematic constraints, it is nonholonomic.

Holonomic constraints are those where the conditions of constraint are expressed as equations connecting the spatial coordinates and time only and do not require higher-order derivatives such as velocity or acceleration in the following form:

$$f\left(r_1, r_2, r_3, \ldots r_n, t\right) = 0 \qquad (2.11)$$

For example, in a rigid body, the distance between any two particles of the body remains constant during motion, which is expressible as:

$$\left| r_i - r_j \right|^2 = c_{ij}^2 \qquad (2.12)$$

where c_{ij} is the distance between the particles i and j with position vectors r_i and r_j. Another example could be of a beaded particle sliding on a circular wire of radius a in the xy plane. In this case, the equation of constraints is:

$$x^2 + y^2 = a^2 \text{ which can be expressed in differential form as } xdx + ydy = 0. \qquad (2.13)$$

A nonholonomic system is one in which there is a continuous, closed circuit of the governing variables. These governing variables can transform the system from any given state to any other state. Since the system's final state depends on its trajectory's intermediate values through parameter space, a conservative potential function cannot represent the system as can, for example, the inverse square law of the gravitational force. In this case, the coordinates are either constrained by inequalities or by nonintegrable differentials.

Gas molecules in a spherical container of radius R are one example of nonholonomic systems. If r_i is the ith molecule's position vector, then the constraint inequality can be expressed as:

$$x_i^2 + y_i^2 + z_i^2 \leq R^2 \qquad (2.14)$$

The center of the sphere here is the origin of the coordinate system. In nonholonomic constraints, the constraints are expressed as relationships between the velocities of the system's particles as:

$$f(x_1, x_2, \ldots, \dot{x}_1, \dot{x}_2, \ldots, t) = 0 \qquad (2.15)$$

If integrating these equations of nonholonomic constraints can define relationships among the coordinates, then the constraints become holonomic. Constraints are additionally categorized as scleronomous and rheonomous. A scleronomous constraint is a time-independent one, whereas a rheonomous constraint includes time. An example of a scleronomously constrained system is a pendulum with an inextensible string of length l_0 that is described by Equation 2.16:

$$x^2 + y^2 = l_0^2 \qquad (2.16)$$

A pendulum with an extensible string is rheonomous; the condition of constraint is:

$$x^2 + y^2 = l^2(t) \qquad (2.17)$$

where $l(t)$ is the length of the string at time t. Constraints add two types of technical issues in solving mechanical problems. The r_i coordinates are no longer independent because constraint equations relate them. This problem is overcome in the case of holonomic constraints by the use of generalized coordinates. The second problem is because the forces of the constraints cannot be clearly defined. These are among the problem's unknowns, which must be derived from the solution. This difficulty can be overcome if the problem is formulated in the Lagrangian form, in which the restriction forces do not appear.

2.2.2 Degrees of Freedom and Generalized Coordinates

The number of independent ways a mechanical device can move without infringing any constraints is called the number of system's degrees of freedom. It is the minimum number of possible coordinates needed to define the system accurately. It has three degrees of freedom when a particle travels through space. If it is constrained to travel along a curve in space, it has just one degree of freedom, while if it moves in a plane, it has two degrees of freedom. In classical mechanics, the motion of a particle to be described requires three space coordinates and one time coordinate; thus, to fully describe the motion of the particle, one has to make use of the three position coordinates (x, y, z) and the rate of change of these coordinates with respect to the time domain, i.e. $(\dot{x}, \dot{y}, \dot{z})$; thus the particle has six degrees of freedom.

Let "s" be a mechanical system that is geometrically constrained. Then the number of generalized coordinates needed to define configuration of s is called degrees of freedom of s. For an N particle system, free from constraints, we need a total of $3N$ independent coordinates to define its configuration fully. Let there be k constraints in the form $f_s(r_1, r_2, ..., r_N, t) = 0$, $s = 1, 2, 3, ..., k$ acting on the system. Now there are only $3N - k$ independent coordinates or degrees of freedom in the system. These $3N - k$ independent coordinates are called the generalized coordinates and are defined by the variables $q_1, q_2, q_3, ..., q_{3N-k}$. The old coordinates $r_1, r_2, ..., r_N$ can be expressed as:

$$r_1 = r_1(q_1, q_2, ..., q_{3N-k}, t), \, r_2 = r_2(q_1, q_2, ..., q_{3N-k}, t), \, ...,$$

$$r_N = r_N(q_1, q_2, ..., q_{3N-k}, t) \tag{2.18}$$

In analogous with Cartesian coordinates, time derivatives $\dot{q}_1, \dot{q}_2, \dot{q}_3, ...$ are defined as generalized velocities. We have shown that a system's entire structure can be determined by $n = 3N - k$ independent generalized coordinates $q_1, q_2, q_3, ..., q_n$. To consider the n q's as the coordinates of a point in an n-dimensional space is useful. This n-dimensional space is called the configuration space, with a coordinate representing each dimension. Since generalized coordinates are not necessarily coordinates of position, configuration space is not necessarily related to the three-dimensional physical space. The motion path also does not necessarily resemble the actual particle spatial direction.

2.2.3 Virtual Work and Lagrange's Equations

A virtual displacement, denoted by δr_i, refers to an imaginary, infinitesimal, instantaneous displacement of the coordinate compatible with the constraints. It is distinct from the actual device displacement δr_i that occurs in a time interval δt. Since displacement is instantaneous, it is called virtual. Because there is no real system motion, the work performed in such a simulated displacement by the forces of constraint is zero. Consider an equilibrated scleronomic array of n particles. Let F_i be the force that acts upon the ith particle. The force F_i is a vector sum of the force applied F_i^e and the forces of the constraints f_i externally. That is, $F_i = F_i^e + f_i$, and the virtual work performed is then given as:

$$\delta W_i = F_i.\delta r_i = \left(F_i^e + f_i\right).\delta r_i \tag{2.19}$$

If the system is in equilibrium, the total force must be zero on each particle: $F_i = 0$ on all i. Hence the dot product $F_i.\delta r_i$ is zero, too. The sum of the above vanishing products is the total virtual work performed on the system δW:

$$\delta W = \sum_{i=1}^{n} \delta W_i = \sum_{i=1}^{n} \left(F_i^e + f_i\right).\delta r_i = \sum_{i=1}^{n} F_i^e.\delta r_i + \sum_{i=1}^{n} f_i.\delta r_i = 0 \tag{2.20}$$

The work done by forces of constraint under a virtual displacement is zero; thus, Equation 2.20 reduces to:

$$\delta W = \sum_{i=1}^{n} \delta W_i = \sum_{i=1}^{n} F_i^e.\delta r_i = 0 \tag{2.21}$$

Equation 2.21 describes the concept of virtual work, which is stated as follows: In the n-particle system, the total work done by external forces when virtual displacements are performed is called virtual work, and the total virtual work performed is zero. The coefficients δr_i in Equation 2.21 can no longer be set at zero since they are not independent. It is important to note here that the virtual work concept deals only with statics. d'Alembert suggested a principle that involves the general motion of the system. In the n-particle system, then by Newton's law, $F_i = \dot{p}_i$ or $F_i - \dot{p}_i = 0$. This equation means that the ith particle in the system would be in equilibrium under a force equivalent to the actual force plus a "reversed effective force" $-\dot{p}_i$, as called by d'Alembert. Thus from Equations 2.19 and 2.20, we get:

$$\sum_{i=1}^{n} \left(F_i^e + f_i - \dot{p}_i\right).\delta r_i = 0 \tag{2.22}$$

Considering the virtual work performed by the forces of constraints to be zero, Equation 2.22 reduces to

$$\sum_{i=1}^{n}\left(F_i^e - \dot{p}_i\right).\delta r_i = 0 \tag{2.23}$$

Generalizing this we can write:

$$\sum_{i=1}^{n}\left(F_i - \dot{p}_i\right).\delta r_i = 0 \tag{2.24}$$

Lagrange used the d'Alembert theory to derive equations of motion, now known as Lagrange equations. Lagrange assumed that the virtual displacements δr_i in Equation 2.24 are not independent but involve virtual displacement of independent generalized coordinates. Assuming a system of N particles positioned at r_1, r_2,..., r_N, with k holonomic constraint equations, this system will have $n = 3N - k$ generalized coordinates $q_1, q_2, q_3,..., q_n$. The transformation from r variables to q variables is given by $r_i = r_i(q_1, q_2, ..., q_{3N-k}, t)$. Since virtual displacement is instantaneous and does not involve time, the virtual displacement of generalized coordinates can be written as:

$$\delta r_i = \frac{\partial r_i}{\partial q_1}\delta q_1 + \frac{\partial r_i}{\partial q_2}\delta q_2 +...+ \frac{\partial r_i}{\partial q_n}\delta q_n = \sum_{j}\frac{\partial r_i}{\partial q_j}\delta q \tag{2.25}$$

$$\dot{r}_i = \frac{\partial r_i}{\partial q_1}.\dot{q}_1 + \frac{\partial r_i}{\partial q_2}.\dot{q}_2 +...+ \frac{\partial r_i}{\partial q_n}.\dot{q}_n + \frac{\partial r_i}{\partial t} = \sum_{j}\frac{\partial r_i}{\partial q_j}.\dot{q}_j + \frac{\partial r_i}{\partial t} \tag{2.26}$$

From Equations 2.28, 2.29, and 2.30, we can write:

$$\sum_{i=1}^{n}F_i.\delta r_i = \sum_{j}\left(\sum_{i}\frac{\partial r_i}{\partial q_j}\right).\delta q_j = \sum_{j}Q_j.\delta q_j \tag{2.27}$$

The Q_j is the jth component of the generalized force Q. Considering the second portion of Equation 2.28:

$$\sum_i \dot{p}_i . \delta r_i = \sum_j \left(\sum_i \frac{d}{dt} \left(m_i \dot{r}_i \frac{\partial r_i}{\partial q_j} \right) - \sum_i m_i \dot{r}_i \frac{d}{dt} \left(\frac{\partial r_i}{\partial q_j} \right) \right) \delta q_j$$

$$= \frac{d}{dt} \frac{\partial}{\partial \dot{q}_j} \left(\sum_i \frac{1}{2} m_i v_i^2 \right) - \frac{\partial}{\partial q_j} \left(\sum_i \frac{1}{2} m_i v_i^2 \right) = \sum_{j=1}^n \left(\frac{d}{dt} \frac{\partial K}{\partial \dot{q}_j} - \frac{\partial K}{\partial q_j} \right)$$

$$(2.28)$$

Substituting Equations 2.32 and 2.31 in Equation 2.28, we get:

$$\sum_{j=1}^n \left(\frac{d}{dt} \frac{\partial K}{\partial \dot{q}_j} - \frac{\partial K}{\partial q_j} - Q_j \right) \delta q_j = 0 \qquad (2.29)$$

Considering

$$F_i = -\nabla_i U \text{ or } Q_j = -\sum_i \nabla_i U \frac{\partial r_i}{\partial q_j} = -\frac{\partial U}{\partial q_j} \qquad (2.30)$$

Equation 2.29 can be rewritten as $\dfrac{d}{dt} \dfrac{\partial K}{\partial \dot{q}_j} - \dfrac{\partial K}{\partial q_j} + \dfrac{\partial U}{\partial q_j} = 0$, and if the

potential U is a function of position only, i.e., $\dfrac{\partial U}{\partial \dot{q}_j} = 0$, we can write:

$$\frac{d}{dt} \frac{\partial}{\partial \dot{q}_j} (K - U) - \frac{\partial}{\partial q_j} (K - U) = 0 \qquad (2.31)$$

This introduces the Lagrangian function $L(q, \dot{q}, t) = K(q, \dot{q}, t) - U(q)$, and

thus Equation 2.31 can be written as $\dfrac{d}{dt} \dfrac{\partial L}{\partial \dot{q}_j} - \dfrac{\partial L}{\partial q_j} = 0$ for all $j = 1, 2, 3, ..., n$.

Those n equations are known as Lagrange's equations, one for each independent generalized coordinate. These constitute a set of n differential equations of the second order for n unknown functions $q_j(t)$, and the general solution includes $2n$ constants of integration.

For specific systems, the forces acting are not conservative, say where one component can be obtained from potential, and the other is dissipative. For such cases, it is possible to write the Lagrange equations as:

$$\frac{d}{dt} \frac{\partial L}{\partial \dot{q}_j} - \frac{\partial L}{\partial q_j} = Q_j' \qquad (2.32)$$

where the force Q_j' does not arise from the potential.

2.3 HAMILTONIAN MECHANICS

Lagrangian formulation describes a system of differential equations of second order, written for a series of variables representing the location of a physical system of interest. The Hamiltonian theorem implies an analogous definition of the system's position and velocity in terms of first-order equations written for independent variables. Hamiltonian equations can be derived from Lagrangian ones by successive application, in a theory of differential equations, of two well-known procedures: reduction of order and change of variables. Both procedures are intended to obtain a system of equations equal to a given system.

In Lagrangian formalism, generalized coordinates (q_i's) and generalized velocities ($\dot{q}_i$'s) are used as independent coordinates to formulate complex issues that result in linear differential equations of second order. In Hamilton's formalism, generalized coordinates and generalized momenta (p_i's) are essential variables for problem formulation. The formulation is based primarily on the system's Hamiltonian function, a combination of system q_i's and p_i's. Mathematically, the resulting linear differential equations in the first order are simpler to manage. The formalism of Hamilton also serves as the base for further advances such as quantum mechanics. The Hamiltonian function for a system is defined as:

$$H = \sum_i p_i \dot{q}_i - L(q, \dot{q}, t) \tag{2.33}$$

where q stands for the generalized n coordinates $q_1, q_2, q_3, ..., q_n$ and given that $p_i = \dfrac{\partial L}{\partial \dot{q}_j}$, we can write:

$$H = H(p, q, t) \quad q = q_1, q_2, q_3, ..., q_n \text{ and } p = p_1, p_2, p_3, ..., p_n, \tag{2.34}$$

meaning H is expressed as a function of generalized coordinates, generalized momenta, and time. The configuration space in Lagrangian formalism is spanned by the n generalized coordinates. The q's and p's are viewed in the same way here, and the space involved is called phase space. It is a space of 2n variables $q_1, q_2, q_3, ..., q_n$ and $p_1, p_2, p_3, ..., p_n$. Each space point represents both the position and the momentum of each particle.

Hamiltonians of the system and variational principles help derive Hamilton's equations of motion. Differentiating Equation 2.33, we get:

$$dH = \sum_i \left(p_i.d\dot{q}_i + \dot{q}_i.dp_i - \frac{\partial L}{\partial q_i}dq_i - \frac{\partial L}{\partial \dot{q}_i}d\dot{q}_i \right) - \frac{\partial L}{\partial t}dt$$

$$= \sum_i \left(\dot{q}_i.dp_i - \frac{\partial L}{\partial q_i}dq_i \right) - \frac{\partial L}{\partial t}dt \tag{2.35}$$

Differentiating 2.34 we get:

$$dH = \sum_i \left(\frac{\partial H}{\partial q_i}dq_i + \frac{\partial H}{\partial p_i}dp_i \right) + \frac{\partial H}{\partial t}dt \tag{2.36}$$

Comparing Equations 2.35 and 2.36, we obtain:

$$\dot{q}_i = \frac{\partial H}{\partial p_i} \, , \, \dot{p}_i = \frac{\partial H}{\partial q_i} \ \& \ \frac{\partial H}{\partial t} = -\frac{\partial L}{\partial t} \tag{2.37}$$

Expressions given in Equation 2.37 are Hamilton's equations of motion. These are sometimes called canonical equations of motion. They constitute a set of differential equations of first order $2n$, which replace Lagrange's differential equations of second order n. Hamilton's equations refer to traditional, holonomic structures. If one of the forces acting on the system is nonconservative, by replacing $\frac{\partial L}{\partial \dot{q}_j}$ with p_i, the Lagrange equation from Equation 2.32 becomes:

$$\dot{q}_i = \frac{\partial H}{\partial p_i} \ \text{and} \ \dot{p}_i = -\frac{\partial H}{\partial q_i} + Q_i \tag{2.38}$$

From Equation 2.37, since $\sum_i p_i \dot{q}_i = \sum_i mv_i^2 = 2K$, the Hamiltonian function can be redefined as $\mathbf{H} = \mathbf{K} + \mathbf{U}$. Therefore, the Hamiltonian function can explain particle dynamics in the system.

The work performed along any closed path is not zero for a nonconservative force, like a particle that is subject to frictional forces. The force principle was developed to ensure object motion can be defined as a simple cause and effect mechanism. It is presumed that a force field exists in three-dimensional space and is represented mathematically as a continuous, integrable field of vectors, $\mathbf{F(r)}$. If time is also continuous and integrable, it is convenient to partition conservative force field energy between a kinetic and a potential term and preserve total energy.

You can find a differential equation that describes the object's dynamics by only representing the total energy as a function or Hamiltonian, $H = K + U$. Integrating the differential equation of motion gives the object's trajectory as it passes through space. These concepts are compelling in practice and can be applied to many problems concerning macroscopic objects' motion.

2.4 CLASSICAL FIELD THEORY

The classical field originated in the nineteenth century when scientists established the proper degrees of freedom for electromagnetic interaction, and the theory was later generalized. A classical field, in real terms, is a dynamical system with an infinite number of degrees of freedom defined by spatial position. A classical field, in mathematical terms, is a section of some fiber bundle that obeys some partial differential equations. Of course, the story has much more to it, but this is a decent first pass at the concept. In contrast, a mechanical system is a dynamic system described by ordinary differential equations with a finite number of degrees of freedom. The ultimate description here is through quantum field theory. Still, sometimes the classical approximation has widespread macroscopic validity (e.g. Maxwell theory), or the classical approximation can help understand the theoretical structure.

Classical fields occur naturally in physics when nonrigid extended objects such as water bodies, elastic solids, and parts of the atmosphere are represented using a classical continuum approximation to their structure. Another place where classical fields appear is a simple definition of matter and its interactions.

2.4.1 Nonrelativistic Field Theory

2.4.1.1 Gravitational Field

Many of the physical fields that are the simplest are vector force fields. The first field theory of gravity was Newton's law of gravitation, in which an inverse square law defines the interaction between two masses. This theory has proven very helpful in predicting planets moving around the Sun. From Equation 2.10, we get:

$$g = G\frac{M}{r^2} \text{ or in general in vector form, we can write } g(r) = -G\frac{M}{r^2}\hat{r} \quad (2.39)$$

where $\hat{r}$ is the position unit vector and the negative sign indicates that the gravity always points toward the Earth. The scientific discovery that inertial mass and gravitational mass are equal with unprecedented precision amounts contributes to the definition of the gravitational force's strength equivalent to the particle's acceleration. This discovery is the starting point for the equivalence principle, which contributes to general relativity. For an independent collection of masses M_i, with the center of mass located at r_i, the gravitational field is given as:

$$g(r) = -G\sum_i \frac{M_i\left(r - r_i\right)}{\left|r - r_i\right|^3} \quad (2.40)$$

If instead of the collection of masses, we have a continuous mass distribution $\rho(X)$, the summation in Equation 2.40 is replaced with an integral over volume, and we have:

$$g(r) = -G\iiint_V \frac{\rho(X)d^3X\left(r - X\right)}{\left|r - X\right|^3} \quad (2.41)$$

Considering Gauss's law for gravity, the gravitational field $g(r)$ can be described as a gradient of a gravitational potential $\Phi(r)$ as follows:

$$g(r) = -\nabla\Phi(r) \quad (2.42)$$

2.4.1.2 Charged Particle in an Electrical Field

A charged test particle with charge q encounters a force F dependent on its charge alone. We may define the electric field E, so that $F = qE$. Where E is the electric field due to a single charged particle, using Coulomb's Law, this field is defined as follows:

$$E(r) = \frac{1}{4\pi\varepsilon_0}\frac{q}{r^2}\hat{r} \quad (2.43)$$

Since the electrical field is conservative, it can also be expressed as a gradient of the electrical potential $V(r)$, a scalar quantity. So

$$E(r) = -\nabla V(r) \quad (2.44)$$

$$\text{Since } F = qE(r) = -q\nabla V(r) = ma \qquad (2.44a)$$

Equation 2.44a can be used to describe the motion of a test charge particle in an electrical field.

2.4.1.3 Charged Particle in a Magnetic Field

Imagine a charged particle in a uniform magnetic field B, with a charge q, mass m, and velocity v. The force which the charge experiences is given by $F = qv \times B$. The equation of motion is written as:

$$m\frac{dv}{dt} = qv \times B \qquad (2.45)$$

Taking the dot product with v on both sides, the right side of the equation becomes zero, thus providing us with the equation:

$$v.m\frac{dv}{dt} = 0 \text{ or}$$

$$\frac{d}{dt}\left(\frac{1}{2}mv^2\right) = 0 \text{ or } K = \text{constant} \qquad (2.46)$$

That is to say that the kinetic energy of a charged particle in a magnetic field remains constant. We can resolve the arbitrary velocity v of the charged particle into two components: one $v_{\parallel}$ parallel to the direction of the magnetic field and the other $v_{\perp}$ that is perpendicular to the magnetic field; the equation of motions in Equation 2.45 can be rewritten as:

$$m\frac{d}{dt}(v_{\parallel} + v_{\perp}) = q(v_{\parallel} + v_{\perp}) \times B \qquad (2.47)$$

Equation (2.9) splits into two equations, one of which describes the motion of the particle parallel to the field and the other describes the motion perpendicular to the magnetic field as:

$$m\frac{dv_{\parallel}}{dt} = 0 \text{ and } m\frac{dv_{\perp}}{dt} = q(v \times B) \qquad (2.48)$$

The set of expressions in Equation 2.48 indicates that the velocity component parallel to the magnetic field $v_{\parallel}$ is constant. Simultaneously, the

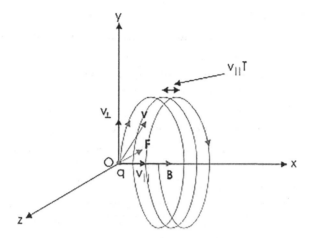

FIGURE 2.2   Charged particle motion in a uniform magnetic field.

acceleration expressed by the quantity $\dfrac{dv}{dt}$ always remains perpendicular to both **v** and **B**, making the particle move in a circle. The total effect of Equation (2.48) is that the particle moves in a helical motion through the magnetic field, as shown in Figure 2.2.

2.4.1.4 Electrodynamics

Generally, when, instead of a point charge, we have a charge density $\rho(r, t)$ and a current density $J(r, t)$, there will be both an electrical and a magnetic field that varies in space and time. The Lorentz force then gives the force **F** experienced by a charge q moving with velocity v in an electromagnetic field, where the electric intensity **E** and magnetic induction **B** are obtainable from the vector potential $A(r, t)$ and scalar potential $V(r, t)$. Maxwell's equations calculate a series of differential equations that directly relate **E** and **B** to the density of electric charge $\rho(r, t)$ (charge per unit volume) and current density (electric current per unit area) $J(r, t)$, which are expressed as:

$$E = -\nabla V - \frac{\partial A}{\partial t} \text{ and } B = \nabla \times A \tag{2.49}$$

Thus, the Lorentz force becomes:

$$F = -q\left[-\nabla V - \frac{\partial A}{\partial t}(v \times (\nabla \times A))\right] \tag{2.50}$$

where q is a test charge placed in the field that experiences the force.

2.4.2 Relativistic Field Theory

Hamilton's variational theory and the Lagrangian mechanics upon which the classical field theory rests are compelling in their application to mechanical systems with a finite number of degrees of freedom. Hamilton's theory characterizes the physically realizable orbits as the essential elements of the fundamental operation within the set of all possible orbits. While not an observable function on its own, the Lagrangian function is not only useful in deriving the equations of motion but also a valuable tool for defining the theory's symmetries and building up the corresponding conserved quantities. The variation and Lagrangian mechanics principle can be generalized to dynamic systems with infinitely uncontrolled many degrees of freedom. The Lagrangian density replaces the Lagrangian function; the coordinates (generalized) are replaced by time- and space-dependent fields. Since we have four independent components as independent fields, we have four equations or one four-vector equation known as the Euler–Lagrange equations. The Euler–Lagrange equations are equations of motion for these fields that derive back to Maxwell's equation based on variational principle.

2.4.2.1 The Action Principle

Let us consider a moving particle along a curve shown in Figure 2.3, and let *a* and *b* be the boundaries along the time axis. Consider all possible trajectories that would minimize the action "*S*" (a move like finding the shortest distance in space between two points). To construct a generalized field theory, consider a four-dimensional box with three space and one time dimensions. The field theory problem tries to find the field everywhere inside the box, provided the field's values on the boundary of the box are known. We need to find a function that would minimize the action "*S*."

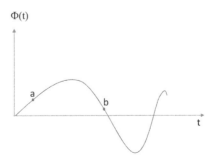

FIGURE 2.3   Nonrelativistic trajectory of a particle in time domain.

In field theory, the usual generalization is to create an event as a sum over tiny space–time cells, an integral part of the space–time frame.

$$S = \int L \, dt \, dx \, dy \, dz \qquad (2.51)$$

where L is a Lagrangian, but from relativity, we know that these four coordinates are considered on an equal basis, each being part of space–time. We obscure the distinction between space and time by giving them identical names – we call them x^i – where the index i runs through all four coordinates and writing the preceding integral from 2.51 as

$$S = \int L \, d^4 x \qquad (2.52)$$

2.4.2.2 Field Theory for Scalar Field

Since the Lagrangian L for a field is distributed over space and time, it is sometimes named Lagrange density. The Lagrangian is a function of space coordinates and velocities. In the notation we used, the Lagrangian is based on a scalar field Φ and partial derivatives of Φ relative to all coordinates: $\Phi, \dfrac{\partial \Phi}{\partial t}, \dfrac{\partial \Phi}{\partial x}, \dfrac{\partial \Phi}{\partial y}, \dfrac{\partial \Phi}{\partial z}$. Thus, we can write:

$$S = \int \left(\Phi, \frac{\partial}{\partial x^i} \right) d^4 x \qquad (2.53)$$

As in ordinary classical mechanics, equations of motion are obtained by wiggling Φ and reducing action S. The Euler–Lagrange equations take the following form for the motion of a particle:

$$\frac{d}{dt} \left(\frac{\partial L}{\partial \left(\frac{\partial \Phi}{\partial t} \right)} \right) - \frac{\partial L}{\partial \Phi} = 0 \qquad (2.54)$$

How will the Euler–Lagrange equations change for the case of multidimensional space–time? We need to adjust Equation 2.54 to use all four space–time directions. For the first term, already reflecting the time domain, we need to take a sum of terms, one for every direction of space–time transforming Equation 2.54 to the following equation:

$$\sum_i \frac{\partial}{\partial x^i}\left(\frac{\partial L}{\partial\left(\frac{\partial \Phi}{\partial x^i}\right)}\right) - \frac{\partial L}{\partial \Phi} = 0 \tag{2.55}$$

Equation 2.55 is the Euler–Lagrange equation for a single scalar field. These equations are closely related to wave equations that characterize wavelike oscillations of Φ. There can be more than a single degree of freedom, as is common in particle mechanics. Where field theory is concerned, this would mean more than one field, so let's presume that there are two fields Φ and X. The action depends on both fields and their derivatives, and there will be an Euler–Lagrange equation for each field:

$$\sum_i \frac{\partial}{\partial x^i}\left(\frac{\partial L}{\partial\left(\frac{\partial \Phi}{\partial x^i}\right)}\right) - \frac{\partial L}{\partial \Phi} = 0 \ \text{and} \ \sum_i \frac{\partial}{\partial x^i}\left(\frac{\partial L}{\partial\left(\frac{\partial x}{\partial x^i}\right)}\right) - \frac{\partial L}{\partial X} = 0 \tag{2.56}$$

More generally, if there are many scalar fields like Φ, each of them will be linked with an Euler–Lagrange equation. A vector field would have additional components that will lead to creating a new Euler–Lagrange equation for each component.

2.4.2.3 Generalization of Field Theory

The nonrelativistic Lagrangian equation contains a kinetic energy term proportional to $\frac{1}{2}\left(\frac{\partial}{\partial t}\right)^2$ and a potential energy term $-U(\Phi)$. We might conclude that generalization to a theory of the field would look similar. Still, with the term kinetic energy having space as well as time derivatives, we have the following expression:

$$L = \frac{1}{2}\left[\left(\frac{\partial \Phi}{\partial t}\right)^2 + \left(\frac{\partial \Phi}{\partial x}\right)^2 + \left(\frac{\partial \Phi}{\partial y}\right)^2 + \left(\frac{\partial \Phi}{\partial z}\right)^2\right] - U(\Phi). \tag{2.57}$$

Space coordinates and time are not symmetric with each other. From relativity, the expression for proper time shows that there is a difference of sign between time and space, i.e.

$$d\tau^2 = dt^2 - dx^2 - dy^2 - dz^2$$

It means we replace the sum of squares (of derivatives) with the difference of squares and obtain:

$$L = \frac{1}{2}\left[\left(\frac{\partial \Phi}{\partial t}\right)^2 - \left(\frac{\partial \Phi}{\partial x}\right)^2 - \left(\frac{\partial \Phi}{\partial y}\right)^2 - \left(\frac{\partial \Phi}{\partial z}\right)^2\right] - U(\Phi) \qquad (2.58)$$

This modified Lagrangian is the Lagrangian for field theory. The function $U(\Phi)$ is called the field potential. This function appears to be like a particle's potential energy. More specifically, at each point of space, it is an energy density (energy per unit volume) that depends on the field value at that point. The function $U(\Phi)$ depends on the context and is deduced from the experiment.

2.5 MAXWELL AND BOLTZMANN EQUILIBRIUM STATISTICS

In 1859, Clark Maxwell introduced the kinetic theory of gases, which Ludwig Boltzmann later updated to explain energy transfer between molecules. Today the combined principles are known as the Maxwell–Boltzmann distribution. The Maxwell–Boltzmann statistics define particles' statistical distribution in thermal equilibrium over different energy states, which sheds light on the microstate. The Maxwell–Boltzmann statistics are applicable when the temperature is high enough and the density is small enough to exclude the quantum effects. Suppose we have an equivalent N point particle gas in a volume V container with the same properties but can be separated from each other, as shown in Figure 2.4. When we say "gas," we think the particles don't interfere with each other. All these particles travel in the container with high velocity in random directions and thus possess different energy. Therefore, the Maxwell–Boltzmann distribution is a function that describes how many particles have some energy in the container.

Maxwell made four key assumptions to develop his distribution models: (1) the diameters of the particles (molecules) are much smaller than the distances between them; (2) the molecules move between collisions

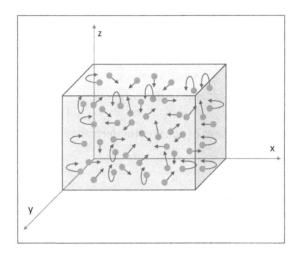

FIGURE 2.4 Identical gas molecules (particles) in a box moving freely and randomly.

without interacting with each other following Newton's first law of motion (i.e. in a straight line at a constant speed); (3) at the initial stages, the positions and velocities of gas molecules are random; and (4) the energy remains conserved during the collision event. The critical assumption is to consider free classical particles in thermal equilibrium at temperature T for deriving the Maxwell–Boltzmann distribution function. Maxwell–Boltzmann's general model of distribution then states that the probability of finding a particle with energy E is proportional to the Boltzmann factor $e^{-\frac{E}{kT}}$ where k is the Boltzmann's constant. If the particle's kinetic energy is $E = \frac{1}{2}mv^2$ where m is the mass and v the velocity, then the normalized distribution of probability for a single particle is given as:

$$P(v) = 4\pi \left(\frac{m}{2\pi kT}\right)^{\frac{3}{2}} v^2 e^{-\frac{mv^2}{2kT}} \tag{2.59}$$

Equation 2.59 describes the probability distribution of a single particle (molecule), so we can write:

$$\int_{0}^{\infty} P(v)\,dv = 1 \tag{2.60}$$

Here, $P(v)dv$ is the probability for the particle to have a speed in the interval $(v; v + dv)$. The maximum value of $P(v)$ occurs at the most probable speed, $v = v_p = \sqrt{\dfrac{2kT}{m}}$, where the first derivative of the function $P(v)$ is zero, i.e. $P'(v) = 0$. The root-mean-squared speed, defined by:

$$v_{rms} = \sqrt{\overline{v^2}} = \sqrt{\frac{3kT}{m}} \tag{2.61}$$

Here,

$$\overline{v^2} = \int_0^\infty P(v)v^2 dv \tag{2.62}$$

Figure 2.5 shows the plot of Maxwell–Boltzmann's probability distribution function against the speed of the particles. It is worth trying to intuitively grasp why $P(v)$ has a limit and falls to zero for both high and low speeds. The high-speed tail, on the right-hand side of Figure 2.5, is easier to understand. There will often be a few particles with speeds above the average with a variety of potential particle speeds during thermal equilibrium. However, the $e^{-\frac{mv^2}{2kT}}$ element in Equation 2.59 quickly suppresses the chance of the velocity being several times greater than v_p. This suppression happens because high-speed particles collide with other particles easily and frequently and usually lose energy in the collision.

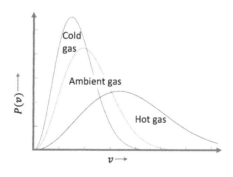

FIGURE 2.5   The Maxwell–Boltzmann probability distribution function in terms of the speeds.

On the other hand, when surrounded by other fast-moving particles, it is challenging for a particle to remain nearly at rest for a long time. The faster particles will strike the slow particles, and usually, kinetic energy transfer will occur in the collision. The Maxwell–Boltzmann probability equations have proven to be extremely useful in studying many natural physical phenomena, especially in materials science and chemistry. The Maxwell–Boltzmann distribution has also found its way into the machine learning community, inspiring a class of mathematical models known as "energy-based" models.

CHAPTER 3

An Overview of Quantum Mechanics

CLASSICAL MECHANICS HAVE BEEN continuously developed since Newton's time and applied to an ever-widening variety of complex systems, including interaction with matter and the electromagnetic field. The philosophy of minimal action governs classical mechanics. When scientists first observed the behavior of electrons and nuclei, they tried to explain their experimental results in terms of Newtonian classical motions, but, in the end, these attempts failed. We find that these small particles behaved in a way that is not in line with Newton's equations.

The classical measurement theory assumes that an interaction between the system of interest and the measuring device can be made arbitrarily small or accurately compensated. One may talk of an idealized measurement that does not interrupt the system's properties of interest. Nevertheless, in the case of an atomic phenomenon, the interaction between the system and measuring instrument is not arbitrarily minimal. Nor can the interaction-related disruption be accounted for precisely because it is uncontrollable and unpredictable to some degree. A calculation on one property will also result in irreversible changes in the value previously allocated to another feature. To talk of a microscopic system with exact amounts for all its wealth is meaningless. Therefore, atomic physics laws must be represented in a nonclassical language, which is a symbolic representation of microscopic measurement principles.

DOI: 10.1201/9781003206743-3

Let us now consider some experimental data that gave rise to these paradoxes and led to quantum mechanics development. Light exhibiting interference was assumed to be a wave phenomenon; however, later, it was observed that shining light on a metal surface results in electrons' ejection from the metal surface (photoelectric effect). Considering the photoelectric effect based on the wave nature of light would conclude that the ejected electron's energy would depend on the intensity of light rather than the frequency observed in experiments.

Rutherford's scattering experiments led to the postulation that the atoms consist of a positive core nucleus and an electronic satellite; why do these electrons not radiate energy on a continuous spectrum and collapse into the nucleus in a discrete line spectrum? Then, the black body radiation could not account for the frequency distribution of the radiant energy based on light's wave nature.

The observation that electrons and other small light particles show wavelike behavior was significant since all atoms and molecules are made up of these particles. If we want to appreciate the molecules' motions and actions better, we need to explain specific properties for their constituents adequately. The classical Newtonian equations do not include influences that indicate wave properties for freely moving electrons or nuclei in space. The above patterns faced tremendous challenges that led to developing new ways to explain these observations. These efforts resulted in the development of quantum mechanics.

3.1 KINEMATICAL FRAMEWORK

When a specific mass m is given, it is perfectly understandable in our everyday physics to talk about the location and velocity of the center of gravity of that mass. However, the relationship $pq - qp = -ih$ between mass, position, and velocity is assumed to hold in quantum mechanics. Here, h is Planck's constant divided by 2π and $i = \sqrt{-1}$. We have good reason to be suspicious if uncritical use of the terms "position" and "velocity" is made. An inconsistency between the definitions of "position" and "velocity" is possible when one acknowledges that discontinuities are often characteristic of processes occurring in small regions and short periods. For instance, if one assumes the motion of a particle in one dimension, then in continuum theory, one can draw (Figure 3.1a) a line $x(t)$ for the track of the particle (more precisely its center of gravity), whose tangent gives the velocity at any moment. In comparison, in a discontinuity-based theory, there may be a sequence of points at finite separation in place of this

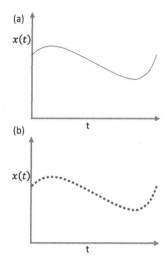

FIGURE 3.1   The trajectory of a particle in classical mechanics (a) vs quantum mechanics (b).

curve (Figure 3.1b). In this case, speaking of one velocity at one position is merely meaningless since two positions and vice versa can only describe one velocity since any single point is correlated with two velocities.

Therefore, the question arises whether it might be possible, by a more detailed study of these cinematic and mechanical principles, to explain the inconsistencies evident up to now in the physical explanations of quantum mechanics and to arrive at a practical understanding of the quantum mechanical formulae. To observe any object's quantum mechanical behavior, one must know the mass of this body and its interactions with specific fields and other objects. Only then can the Hamiltonian function for the quantum mechanical system be written down.

Throughout this discussion, the idea of "the electron's position" seems to be well defined, and only the "size" of the electron needs to be considered. When two high-speed particles strike an electron one after the other within a short time δt at a distance δl, the electron positions described by the two particles exist very close together. From the regularities observed for α-particles, we infer that if the only δt is small enough and particles are selected with sufficiently large velocity, δl can be pushed down to a magnitude of the order of 10^{-12} cm. We mean this when we say the electron is a corpuscle whose radius is no greater than 10^{-12} cm.

Now let us consider the concept of the "path of the electron," described by the series of positions which electron takes one after the other. It is easy to understand, for example, that from our point of view, the commonly used expression "1s orbit of the electron in the hydrogen atom" has no significance. To calculate this 1s "path," we must illuminate the atom with light whose wavelength is considerably less than 10^{-8} cm. However, one single photon of this light is enough to fully eject the electron from its "trajectory" (so that only one point of such a path can be defined). The word "path" has, therefore, no definable significance here.

Conversely, the expected position calculation may be performed on several atoms in 1s state. So the electron's position must have a probability function for a specific state – for example, the 1s state – of the atom, which corresponds to the mean value for the classical orbit, averaged over all phases, can be calculated with arbitrary precision through the measurement. In other words, a numerical value for the position, velocity, and momentum of a particle does not exist. In quantum mechanics, what exists is described by a wave function $\psi(x, y, z, t)$. A visual depiction of an arbitrary wave function is given in Figure 3.2. This function is also known as Schrödinger's wave function.

The physical nature of the wave function is known as "Born's statistical interpretation": darker regions are regions where, if the distance is narrowed down, the particle is more likely to be located. More specifically, if $\vec{r}(x, y, z)$ is a given position, then if such a measurement is attempted, the likelihood of finding the particle inside a small range, of size $d^3\vec{r} = dxdydz$, around that location, is given by:

$$\left|\psi(\vec{r},t)\right|^2 d^3\vec{r} \tag{3.1}$$

Furthermore, if such a position evaluation is performed, the wave function would be affected. After the assessment, the new wave function will be

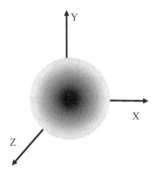

FIGURE 3.2   Visualization of an arbitrary wave function.

limited to the amount to which the location has been narrowed down, but it can expand out again in time if it is allowed to do so later. There must be a particle somewhere. In quantum mechanics, which is demonstrated by the fact that the overall probability of finding the particle, combined in all possible locations, must be 100% (certainty):

$$\int\limits_{all\ \vec{r}} \left|\psi(\vec{r},t)\right|^2 d^3\vec{r} = 1 \qquad (3.2)$$

In other words, normalizing appropriate wave functions, $\psi|\psi = 1$.

3.1.1 Heisenberg's Uncertainty Principle

Unless one wants to be clear about what is to be interpreted by the term "position of the object," for example, the electron (relative to a given frame of reference), then one must identify such experiments with whose aid one aims to determine the "position of the electron"; otherwise, this term has no significance. Let one light up the electron, for instance, and study it under a microscope. Then the maximum achievable precision in position estimation is controlled by the light's wavelength. In theory, however, one can construct, say, a y-ray microscope and, in addition to that, carry position determination with as much precision as one likes. There is an essential aspect of that calculation, the Compton effect. Any detection of scattered light coming from the electron presupposes a photoelectric effect. It can, therefore, also be perceived in such a way that a light quantum reaches the electron, is reflected or scattered, and then, again bent by the microscope lens, creates the photo effect.

The electron undergoes a discontinuous adjustment in momentum as the position is determined – since, at that instance, the photon is scattered by the electron. This shift in momentum is, the more significant, the smaller the wavelength of the light used – that is, the more precise the location determination is. Therefore, when the electron's location is determined, simultaneous momentum may be known up to the magnitudes corresponding to that discontinuous shift. Thus, the more accurately the position is calculated, the less precisely the momentum is established, and vice versa. In this situation, we see a clear physical description of the $pq - qp = -i\hbar$ equation.

Heisenberg postulated uncertainty relationships. Simple particle behavior involves particle localization, whereas simple wave behavior only occurs when the particle has a definite momentum. Heisenberg's understanding of this was that only when the other is untenable, each of these extreme classical definitions is fulfilled. For intermediate circumstances, neither picture is accurate. However, in terms of trajectories, quantum mechanics must be consistent with the description of elementary particles' motion (not just the description of the movement of macroscopic bodies). The response from Heisenberg's uncertainty principle is that one can create states that require a certain amount of momentum and position localization. Furthermore, a particle's motion has a certain similarity to classical movement around trajectories.

Nevertheless, there should be a specific spread in the momentum and the position coordinate so that the amplitudes in momentum and position allow uncertainty relationships to be maintained. The Heisenberg uncertainty postulates are a convenient way to describe the qualitative properties of quantum mechanics. Figure 3.3a is a combination plot of a particle's position x and the respective linear momentum p. Figure 3.3b describes what happens if we force the particle down to try and limit it to one position x: It extends in the direction of momentum. Heisenberg showed that the blue "blob" region could not be compressed to a point according to quantum mechanics. When we try to narrow down a particle's location, we get into trouble with momentum. Conversely, we lose all grip on the position if we try to pin down a specific momentum.

Quantitatively, if an equivalent experiment involving an electron is carried out several times, then electron's position x is calculated in each run of the trial. Simultaneously, the experimental setup is similar in each series; the electron's position measurement does not yield the same result. Let the average position measurement be $\langle x \rangle$; then $(\Delta x)^2 \equiv \langle (x - \langle x \rangle)^2 \rangle$ gives the mean square deviation, which means that the standard deviation is Δx.

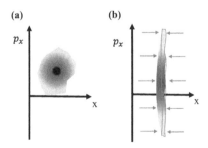

FIGURE 3.3 (a) Combined plot of position and momentum (b) illustrating uncertainty principle.

If Δx is small compared with any length in the experiment, one is more precise in any given run to find the value $x = \langle x \rangle$. If Δx is broad, it is not sure what will yield the value of x. So, the uncertainty in the value of x is called Δx.

Likewise, one can speak of uncertainty in any physically measurable quantity such as momentum, energy, and field. Mathematically, Heisenberg's uncertainty relationship states that: $\Delta x \Delta p \gtrsim \hbar$

If we can say with certainty what the position of a particle is (i.e. $\Delta x = 0$), we are unsure about its momentum (i.e. $\Delta p = \infty$). Heisenberg's uncertainty relationships between momenta and coordinates are generalized to every measurable pair and appear as a product of their commutation relationships. Complementary variables are called measurable parameters that follow uncertainty relation. Examples include position and momentum (x, p), energy and time (E, t), and any two Cartesian coordinates of angular momentum (L_x, L_y).

3.1.2 Quantum Mechanical Operators

The numerical quantities used by the old Newtonian mechanics, position, momentum, and energy are mere "shadows" of what nature describes: operators. The operators mentioned in this section are the fundamental elements of quantum mechanics. There corresponds to a single linear operator for any physical quantity. In particular, the operators $\hat{x}$ and $\hat{p}$, corresponding to a particle's x-coordinate and momentum, fulfill the commutation relationship $[\hat{x}, \hat{p}] = i\hbar$ As the first example, although there is never a mathematically accurate value of a particle's position x, we have an operator $\hat{x}$, which translates the wave function ψ into x:

$$\psi(x,y,z,t) \xrightarrow{\hat{x}} x\psi(x,y,z,t) \tag{3.3}$$

The operators $\hat{y}$ and $\hat{z}$ are defined similarly. In the case of the x-component of a linear momentum $p_x = mv_x$, we have an x-momentum operator that turns ψ into its x-derivative:

$$\psi(x,y,z,t) \xrightarrow{\widehat{p_x} = \frac{\hbar}{i}\frac{\partial}{\partial x}} \frac{\hbar}{i}\psi_x(x,y,z,t) \tag{3.4}$$

where $\hbar$ is Planck's constant and provides an approximation of the domain in which quantum mechanics becomes important. It has classical action dimensions (*energy* × *time*). Classical physics would refer to systems where the action is much higher than $\hbar$. If it had been zero, we would not have had all these issues with quantum mechanics. The blobs will be turning into points. Alas, $\hbar$ is very small but nonzero. It is approximately 10^{-34} kg m²/s. The element i in $\hat{p}_x$ makes the operator Hermitian. All operators that represent our physical macroscopic quantities are Hermitian. Both the $\hat{p}_y$ and $\hat{p}_z$ operators are similarly described as $\hat{p}_x$. Moreover, thus, the kinetic energy operator $\hat{K}$ is given as:

$$\hat{K} = \frac{\hat{p}_x^2 + \hat{p}_y^2 + \hat{p}_z^2}{2m} = -\frac{h^2}{2m}\left(\frac{\partial^2}{\partial x^2} + \frac{\partial^2}{\partial y^2} + \frac{\partial^2}{\partial z^2}\right) = -\frac{h^2}{2m}\nabla^2 \tag{3.5}$$

In quantum mechanics, the relationships between the various operators are necessarily the same as those between the corresponding numerical values in Newtonian physics. Mathematicians call the "Laplacian" the set of second-order derivative operators in the kinetic energy operator $\hat{K}$ and indicate it by ∇^2. According to the Newtonian analogy, the total energy operator indicated by $\hat{H}$ is the sum of the above kinetic energy operator and the potential energy operator $\hat{U}(x, y, z, t)$:

$$\hat{H} = -\frac{\hbar^2}{2m}\nabla^2 + \hat{U} \tag{3.6}$$

The possible results of the measurements q_i of a physical quantity Q are called the eigenvalues for the corresponding quantum mechanical operator $\hat{Q}$. If the state vector $\psi = \sum c_i \varphi_i$ represents the state of the system, then the modulus squared $|c_i|^2$ of the eigenvector φ_i is the probability of obtaining the value q_i. The total energy operator $\hat{H}$ is called the Hamiltonian, and it is essential in solving quantum mechanical problems. Its eigenvalues are indicated by E (for energy), for example, E_1, E_2, E_3,... In particular, the possible values of the energy E_i are obtained by solving the eigenvalue equation:

$$\hat{H}\varphi_i = E_i\varphi_i \tag{3.7}$$

A more complex numbering of the Hamiltonian's eigenvalues and eigenvectors is ideal in certain situations, instead of using a single counter n. For example, for the hydrogen atom electron, there is more than one eigenfunction for each E_n eigenvalue, and additional l and m counters are used to differentiate them. Generally, it is possible to solve the question of eigen-value first and then determine whether to number the solutions later.

We observed the existence of relationships between the physical world and mathematics that are different from those defined in classical physics. In quantum mechanics, physical quantities are related to (non-commuting) operators; operations build the state vectors with these mathematical enti-ties; the input to the physical world is rendered by measurements that produce the corresponding operators' proper values possible outcomes. An example of this two-way relation between formalism and the physical world is as follows: Assume that the framework is established in a specific physical state assigned to the state vector ψ. This assignment is evaluated using specific samples, i.e. measurements of observables Q. We may be familiar with the corresponding eigenvector φ_i and hence the likelihood of obtaining the eigenvalues q_i. This bidirectional relationship between the physical world and formalism is not a simple one to grasp.

The most challenging part of studying quantum mechanics is under-standing how theoretical formalism can be applied to real laboratory phenomena. These applications include almost always formulating over-simplified theoretical models of the actual phenomena, to which quantum formalism can be applied effectively. The best physicists have an excep-tional understanding of what characteristics of the real phenomena are essential and must be described in the theoretical model and what charac-teristics are irrelevant and can be ignored.

The orthodox explanation is that nature involves a mysterious source of random numbers. In terms of eigenfunction, if the wave function ψ before the "measurement" is equal to:

$$\psi = c_1\psi_1 + c_2\psi_2 + c_3\psi_3 + \ldots \tag{3.8}$$

Then, in Einstein's words, this random number generator "throws the dice" and selects one of its eigenfunctions based on the outcome. The wave function will collapse in ψ_1 in $|c_1|^2$, the fraction of the cases on average, and the wave function will collapse in ψ_2 in $|c_2|^2$. The orthodox definition says that the eigenfunction coefficients' square magnitudes give the probabili-ties of the corresponding eigenvalues.

3.1.3 Quantum Mechanical Energy States

Let us consider a one-particle system to be studied, say, an electron, con-fined to the inside of a short, sealed-end pipe. Studying this example in

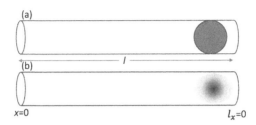

FIGURE 3.4 (a) Classic illustration of a particle in a closed pipe. (b) Quantum mechanics picture of a particle in a closed pipe.

some depth may make it much easier not to get lost in more complex quantum mechanics cases. Figure 3.4a illustrates the structure we want to analyze as it would appear in classical nonquantum physics. A particle bounces in between a pipe's two ends. If friction is not present, the particle will continue to bounce back and forward forever. Friction is a macroscopic phenomenon with no place in the kind of quantum-scale structures we want to study here. Classical physics usually draws the particles it defines as small spheres, as shown in Figure 3.4a.

We need to find the Hamiltonian, the sum of kinetic and potential energy operators, to evaluate the system. As we presume that no forces act on the particle (i.e. before it reaches the pipe ends), the potential energy must be constant and supposed to be zero for practical purposes. The kinetic energy operator then becomes Hamiltonian operator $H = -\dfrac{h^2}{2m}\dfrac{\partial^2}{\partial x^2}$. The next step is to formulate the Hamiltonian problem of eigenvalues (or "time-independent Schrödinger equation"). This problem is always the form of $H\psi = E\psi$. Every nonzero solution of this equation is called an energy eigenfunction, and the corresponding constant E is called the energy eigenvalue.

Substituting the Hamiltonian for the pipe as found in the previous paragraph, the eigenvalue problem is:

$$-\frac{h^2}{2m}\frac{\partial^2 \psi}{\partial x^2} = E\psi \tag{3.9}$$

We are not done yet. We need the so-called "boundary conditions," conditions that tell what is happening at the end of the x range. In this case, the x range ends will be the pipe ends. Now note that the wave function's square magnitude gives the likelihood that the particle can be found. So, the wave function must be zero; anywhere, the particle cannot be located. When the particle is confined to the pipe, the wave function outside the tube is zero, so that the wave function at the ends will be zero. Given these boundary conditions, the solution to differential Equation 3.9 is as follows:

$$E_n = \frac{n^2 h^2 \pi^2}{2m l_x^2} \text{ for } n = 1, 2, 3, \ldots \tag{3.10}$$

There is one more condition that must be satisfied. Each solution must be standardized so that the total probability of finding the particle incorporated is 1 (certainty) among all the possible positions. That calls for:

$$\langle \psi_n | \psi_n \rangle = 1 = \int_{x=0}^{l_x} |C_2|^2 \sin^2\left(\frac{n\pi}{l_x} x\right) dx \tag{3.11}$$

The solution to Equation 3.11 provides:

$$\psi_n = \sqrt{\frac{2}{l_x}} \sin\left(\frac{n\pi}{l_x} x\right) \text{ and } E_n = \frac{n^2 h^2 \pi^2}{2m l_x^2} \text{ for } n = 1, 2, 3, \ldots \tag{3.12}$$

Typically, the energy values are displayed graphically in the form of an "energy continuum," as shown in Figure 3.5. Energy is plotted upward, and the vertical height of each amount of energy shows how much energy it has. The solution counter, or "quantum number," is mentioned to the right of each energy point, n. Classically, the particle's total energy may be of any nonnegative value. Nevertheless, this is not valid according to quantum mechanics: total energy must be one of the amounts shown in

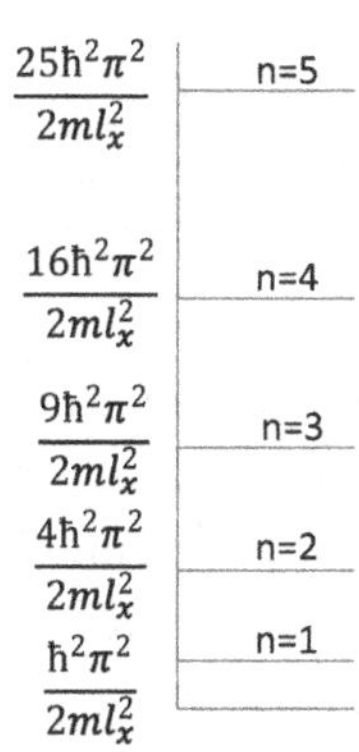

$\dfrac{25\hbar^2\pi^2}{2ml_x^2}$	n=5
$\dfrac{16\hbar^2\pi^2}{2ml_x^2}$	n=4
$\dfrac{9\hbar^2\pi^2}{2ml_x^2}$	n=3
$\dfrac{4\hbar^2\pi^2}{2ml_x^2}$	n=2
$\dfrac{\hbar^2\pi^2}{2ml_x^2}$	n=1

FIGURE 3.5   One-dimensional energy spectrum for a particle in a pipe.

Figure 3.5 of the energy spectrum. It should be noted that one will not know the difference for a macroscopic particle; the distance between the energy levels is macroscopically excellent, as Planck's constant ℏ is so small. However, the discreteness of the energy values will make a significant difference in a quantum-scale system.

The particle will be trapped in the lowest possible energy level at absolute zero temperature, $E_1 = \dfrac{\hbar^2 \pi^2}{2ml_x^2}$, in the spectrum of Figure 3.5. This lowest possible level of energy is called the "ground state." Classically, one assumes that the particle has no kinetic energy at absolute zero, so zero overall energy. However, this is not allowed by quantum mechanics. Heisenberg's theory requires some momentum and kinetic energy to remain even at zero temperature for a confined particle.

The solution to the quantum mechanical problem of a particle represented in Equation 3.12 is for one dimension only, i.e. the x-direction. Real life is three-dimensional; if we presume the pipe has a square cross section, we can solve the problem relatively quickly. There is a way to turn one-dimensional solutions into complete three-dimensional solutions for all three coordinates. This solution is called the principle of "variable separation": Solve each of the three variables x, y, and z separately and then combine the results. The complete three-dimensional problem has its eigenfunctions $x_n y_n z_n$, which are merely products of the one-dimensional ones, and the energy eigenvalues $E_{x_n y_n z_n}$ of the three-dimensional problem are the sum of those of the one-dimensional issues and are given as follows:

$$\psi_{n_x n_y n_z} = \sqrt{\frac{8}{l_x l_y l_z}} \sin\left(\frac{n_x \pi}{l_x} x\right) \sin\left(\frac{n_y \pi}{l_y} y\right) \sin\left(\frac{n_z \pi}{l_z} z\right) \qquad (3.13)$$

$$E_{n_x n_y n_z} = \frac{n_x^2 \hbar^2 \pi^2}{2ml_x^2} + \frac{n_y^2 \hbar^2 \pi^2}{2ml_y^2} + \frac{n_z^2 \hbar^2 \pi^2}{2ml_z^2} \qquad (3.14)$$

3.1.4 Quantum Confinement

In physics, movement usually happens in three dimensions. In classical mechanics, a particle in a pipe will be able to travel in all three dimensions. However, in quantum mechanics, the motion is entirely one-dimensional when the tube gets very small. This matter also has to do with the fact that the amounts of energy are distinct in quantum mechanics. For example, the kinetic energy in the y-direction takes the possible values that will be very large for a narrow pipe in which l_y is small. Indeed, the particle will have the high energy E_{y1} in the y-direction; provided it is in the tube at all, it will have at least that amount of energy. However, it likely will not have enough additional thermal energy to get to the next E_{y2} state. For this, the kinetic energy in the y-direction will be trapped at the lowest E_{y1} level possible.

The argument is not that the particle is not "moving" in the y-direction; literally, E_{y1} is a great deal of kinetic energy. It is all about this energy frozen into one state. The particle has no other energy state to "play in" in the y-direction. If the pipe in the z-direction is also extremely short, the only essential and relevant motion is in the x-direction making nontrivial physics completely one-dimensional. We have produced a "quantum wire."

However, suppose the pipe size in the z-direction is relatively large. In that case, the particle will still have lots of different energy states available in the z-direction, and the motion will be two-dimensional, a "quantum well." On the other hand, if the pipe is small in all three dimensions, we get a zero-dimensional "quantum dot" in which the particle does little until it absorbs a significant amount of energy. It will give readers a rough understanding of all the fascinating things one can do in nanotechnology when one restricts particle movement, mainly electrons, in different directions. One is turning the dimensionality of our standard three-dimensional universe into a lower one. Only quantum mechanics can explain why by making the energy levels discrete rather than continually varying. Moreover, the smaller dimensions' cosmos will have one's topology preference (a triangle, a letter 8, a circle, a cylinder) to make it very interesting.

3.2 DYNAMICS OF QUANTUM MECHANICAL SYSTEMS

Let us consider the example of a particle bound by forces to stay in virtually the same position. It may be defined as objects such as an atom or a molecule in a crystalline solid. Suppose the forces driving the particle back to its nominal location are proportional to the displacement and are directed toward the particle equilibrium location. In that case, we have what is called a harmonic oscillator. This relationship is generally a reasonable approximation for other restricted systems if the distance from the nominal position remains high. We will show the particle's displacement from the nominal position (x, y, z). It is possible to model the forces which hold the particle confined as springs, as sketched in Figure 3.6.

The spring stiffness is defined by the so-called "spring constant" k, which gives the ratio of force to displacement. Note that we are going to assume the three spring stiffnesses are equivalent. The particle vibrates back and

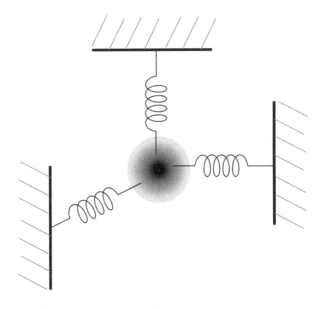

FIGURE 3.6   Molecular harmonic oscillator.

forth around its nominal location with a frequency, according to classical Newtonian physics $\omega = \sqrt{\dfrac{k}{m}}$. This frequency remains a convenient computational quantity in quantum mechanics as well. We must first write down the total Hamiltonian energy to find the oscillating particle's energy levels. In terms of potential energy, the spring in the x-direction carries a possible energy quantity equal to $\dfrac{1}{2}kx^2$, and likewise, those in y- and z-directions. Thus, the Hamiltonian from Equation (3.6) can be written as:

$$\hat{H} = -\frac{\hbar^2}{2m}\nabla^2 + \frac{1}{2}k\left(x^2 + y^2 + z^2\right) \tag{3.15}$$

Next, we need to use the Hamiltonian to find the harmonic oscillator's energy eigenfunctions and eigenvalues. Every eigenfunction ψ and its energy eigenvalue E will satisfy the "time-independent Schrödinger's equation":

$$\hat{H}\psi = \left[-\frac{\hbar^2}{2m}\nabla^2 + \frac{1}{2}k\left(x^2 + y^2 + z^2\right)\right]\psi = E\psi \tag{3.16}$$

The boundary condition is that ψ is zero at a substantial distance from the nominal position. After all, the magnitude of ψ informs us of the relative possibility of locating the particle at that position. Because of the exponentially increasing potential energy, the odds of finding the particle far from the nominal position will be negligible. It is assumed that every eigenfunction is a product of one-dimensional eigenfunctions. Expanding and solving Equation (3.16) reveals that the energy eigenvalue is a combination of three eigenvalues E_x, E_y, and E_z, such that $E = E_x + E_y + E_z$ that can lead to the one-dimensional eigenvalue problem:

$$\left[-\frac{\hbar^2}{2m} \frac{\partial^2}{\partial x^2} + \frac{1}{2} kx^2 \right] \psi_x = E_x \psi_x \tag{3.17}$$

Here, the operator inside the square brackets, call it Hx, only uses the x-related terms in the Hamiltonian. Similar expressions with E_y and E_z can be written down. We can define separate different problems in each of the three variables x, y, and z. It is possible to solve the one-dimensional problem for ψ_x by fundamental means. Like the particle in the previous section's pipe, there are an infinite number of different solutions for E_x and ψ_x. Unlike the particle in the tube, the solutions are numbered here, starting from 0 instead of 1. So E_{x0} and ψ_{x0} are the first eigenvalue and eigenfunction, respectively:

$$E_{x0} = \frac{1}{2} \hbar\omega \qquad \psi_{x0} = h_0(x)$$

$$E_{x1} = \frac{3}{2} \hbar\omega \qquad \psi_{x1} = h_1(x) \tag{3.18}$$

$$E_{x2} = \frac{5}{2} \hbar\omega \qquad \psi_{x2} = h_2(x)$$

Also, the eigenfunctions themselves are not sines like the ones for the particle in the pipe; instead, they take an exponential form with some polynomial multiplier. The general formula for the eigenfunction is given as:

$$h_n = \frac{1}{\left(\pi l^2 \right)^{1/4}} \frac{H_n(\xi)}{\sqrt{n! 2^n}} e^{-\xi^2/2} \text{ for } n = 0,1,2,3\ldots \tag{3.19}$$

where $\omega = \sqrt{\frac{k}{m}}$, $l = \sqrt{\frac{\hbar}{m\omega}}$, $\zeta = \frac{x}{l}$, and H_n is the "Hermite polynomial." However, from this approach, it is the eigenvalues one might want to note. According to the orthodox definition, these are the observable values of total energy in the x-direction (potential energy in the x-direction plus the kinetic energy of motion in the x-direction). They can be represented using the generic expressions:

$$E_{xn_x} = \frac{2n_x + 1}{2}\hbar\omega, \; E_{yn_y} = \frac{2n_y + 1}{2}\hbar\omega \text{ and } E_{zn_z} = \frac{2n_z + 1}{2}\hbar\omega \text{ for } n_x = 0,1,2,3...,n_y$$

$$= 0,1,2,3..., \text{ and } n_z = 0,1,2,3 ...$$

The complete system's total energy E is the sum of E_x, E_y, and E_z. The cumulative total energy value is obtained by nonnegative value for number n_x, combined with nonnegative options for n_y and n_z. Putting these total energy values into the terms for the three partial energies above, we get:

$$E_{n_x n_y n_z} = \frac{2n_x + 2n_y + 2n_z + 3}{2}\hbar\omega \tag{3.20}$$

3.2.1 Energy Eigenvalue for Harmonic Oscillator

For each set of three nonnegative whole numbers n_x, n_y, and n_z, there is one unique energy eigenfunction, or eigenstate, and the corresponding energy eigenvalue. The n_x, n_y, and n_z, "quantum numbers" refer to the numbering scheme of the one-dimensional solutions that make up the complete solution. Eigenvalues are of high physical significance. According to the orthodox understanding, they are the only observable values of total energy, the single energy levels that the oscillator can ever be found to occupy.

As shown in Figure 3.7, the energy levels can be plotted in the form of a so-called "energy spectrum." The energy values are described along the vertical axis, and the sets of quantum numbers n_x, n_y, and n_z are shown on the right side of the diagram. The first point of interest demonstrated by the energy distribution is that no arbitrary value can be inferred by the oscillating particle's energy but only specific distinct values. Of course, that is just like the particle in the pipe, except the energy levels are equally distributed for the harmonic oscillator. The energy value always is an odd multiple of ½$\hbar\omega$. It contradicts the Newtonian notion of any energy level being necessary for a harmonic oscillator. However, because $\hbar$ is extremely small, about 10^{-34}kg cm²/s energy levels are too close together macroscopically. Although the old Newtonian theory is strictly false, it remains an excellent approximation for the macroscopic oscillator.

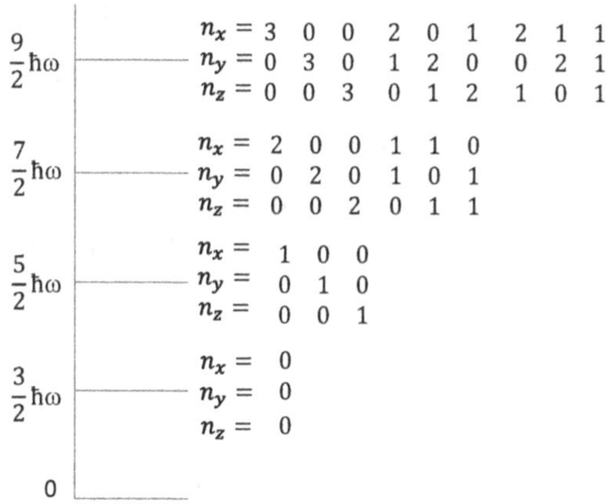

$$\frac{9}{2}\hbar\omega$$

$$
\begin{array}{ccccccccc}
n_x = 3 & 0 & 0 & 2 & 0 & 1 & 2 & 1 & 1 \\
n_y = 0 & 3 & 0 & 1 & 2 & 0 & 0 & 2 & 1 \\
n_z = 0 & 0 & 3 & 0 & 1 & 2 & 1 & 0 & 1
\end{array}
$$

$$\frac{7}{2}\hbar\omega$$

$$
\begin{array}{cccccc}
n_x = 2 & 0 & 0 & 1 & 1 & 0 \\
n_y = 0 & 2 & 0 & 1 & 0 & 1 \\
n_z = 0 & 0 & 2 & 0 & 1 & 1
\end{array}
$$

$$\frac{5}{2}\hbar\omega$$

$$
\begin{array}{ccc}
n_x = 1 & 0 & 0 \\
n_y = 0 & 1 & 0 \\
n_z = 0 & 0 & 1
\end{array}
$$

$$\frac{3}{2}\hbar\omega$$

$$
\begin{array}{c}
n_x = 0 \\
n_y = 0 \\
n_z = 0
\end{array}
$$

$$0$$

FIGURE 3.7 The energy spectrum of the harmonic oscillator.

Note also that the energy levels have no extreme value; however high the particle energy can be in an actual harmonic oscillator, it can never escape. The more it attempts to get free, the stronger the powers that hold it back. It cannot prevail. Another attractive characteristic of the energy continuum is that the lowest possible energy is again nonzero. For $n_x = n_y = n_z = 0$, the lowest energy exists and has the value: $E_{000} = \frac{1}{2}\hbar\omega$. So, even at absolute zero temperature, the particle at its nominal location is not absolutely at rest; it still has $\frac{3}{2}\hbar\omega$ remaining worth of kinetic and potential energy that it will never eliminate. This lowest level of energy is the ground state.

The uncertainty principle can explain the reason the energy can never be zero. The particle will have to be certainly at its nominal position to get the potential energy to be zero. However, the uncertainty principle does not allow for a given position. Also, to get the kinetic energy to be zero, the linear momentum will have to be zero for sure, which is not allowed either by the uncertainty relation. The actual ground state is a compromise between momentum and position uncertainties, which makes the total energy as small as the relationship with Heisenberg's uncertainty principle allows. There is ample momentum instability to hold the particle close to the nominal location, reducing potential energy.

There is enough momentum uncertainty to keep the particle close to the nominal position to minimize potential energy. Still, there is also much uncertainty to keep the momentum small to minimize kinetic energy. Yes, the agreement results in almost equal potential and kinetic energies. Figure 3.7 shows a steadily increasing number of different quantum numbers n_x,

n_y, and n_z, all supplying the energy. Since every collection represents one state, this means multiple states generate the same energy.

3.2.2 Eigenfunction for the Harmonic Oscillator

The particle is in the lowest energy ground state at absolute zero temperature. The eigenfunction representing this state has the lowest possible numbering $n_x = n_y = n_z = 0$ and is equal to:

$$\psi_{000} = h_0(x)h_0(y)h_0(z) \tag{3.21}$$

where function h_0 is obtained from Equation 3.19. Applying the expression for h_0 from Equation 3.19, ground-state properties can be investigated. It shows that the particle is most likely to be located near the ground state's nominal position. The likelihood of finding the particle drops rapidly to zero over a certain distance away from the nominal position. The particle is likely to be detected around a radius, approximately speaking, $l = \sqrt{\hbar/m\omega}$ from the nominal position. It would be a very short distance for a macroscopic oscillator because of the minimal value of ħ. That is very reassuring because we would expect an oscillator to rest at the nominal position, macroscopically. While quantum mechanics is not permitted, the distance l from the nominal position and the energy $\frac{3}{2}\hbar\omega$ are infinitely small.

However, the bad news is that the ground-state probability density does not mimic the classical Newtonian image of a localized particle oscillating back and forth. Besides, the density of probabilities does not depend on time: The chances of finding the particle at any given position are always the same. The probability density is also spherically symmetric; it depends only on the distance from the nominal location and is the same in all angular orientations. One prerequisite to get something that can begin to imitate a Newtonian spring-mass oscillator is that the energy is far above ground level.

Considering the second lowest energy point, three separate energy eigenfunctions ψ_{100}, ψ_{010}, and ψ_{001} attain this energy level. The probability distribution for each of the three takes the form of two different "blobs;" when viewed along the z-direction, Figure 3.8 shows ψ_{100}, and ψ_{010}. One blob covers the other in the case of ψ_{001}, so this function is not seen. These states do not look at all like a Newtonian oscillator, either. Once again, the probability distributions always remain the same. Though a single particle occupies two blobs in each case, the particle will never be trapped in between the blobs on the symmetry axis.

FIGURE 3.8   Wave functions ψ_{100} and ψ_{010}.

3.2.3 Non-Eigen States

It should not be taken for granted that the harmonic oscillator exists only in energy eigenstates. The situation is almost like the reverse. Something that locates the particle's position will create an energy uncertainty. This segment discusses the procedures for dealing with nonenergy eigenstates. First, even if the wave function is not an energy-specific function, it can still be written as a combination of the individual functions:

$$\psi(x, y, z, t) = \sum_{n_x=0}^{\infty} \sum_{n_y=0}^{\infty} \sum_{n_z=0}^{\infty} c_{n_x n_y n_z} \psi_{n_x n_y n_z} \tag{3.22}$$

This possibility stems from a consequence of the Hermitian operators' eigenfunctions' completeness, such as the Hamiltonian.

3.3 QUANTUM STATISTICAL MECHANICS

Thermodynamics and classical statistical physics still lack a commonly accepted and conceptually straightforward foundation despite being very well confirmed by experiments. Quantum mechanics asserts that it is a foundational theory. As such, it should be able to provide us with a microscopic explanation for all the anomalies we find in macroscopic structures. The basic postulates of classical statistical mechanics are now applied to quantum statistical mechanics. In classical mechanics, the variables $\{q, p\}$ characterize a microstate that evolves according to the Hamiltonian equations. A microstate in quantum mechanics is characterized by a wave function $\psi(x, t)$, which develops according to Schrödinger's equation. To start, we describe some of the fundamental definitions.

3.3.1 Expected Value and Standard Deviation

The physical quantities that cannot have value are a surprising result of quantum mechanics. This situation arises if the wave function is not the quantity of interest's eigenfunction. For example, the ground state of the hydrogen atom is not the position operator $\hat{x}$'s eigenfunction, so there is no real value for the electron's x-position. It is impossible to predict what measurement of such a quantity would achieve, as per the orthodox understanding. Nonetheless, it may be known if the same calculation is performed on many systems, which are the same before the calculation is completed. An example would be measurements of x-position on several hydrogen atoms, all of which are in the ground state before measurement. In that case, estimating the sum or "expected value" of all the measures would be straightforward.

Indeed, the expected value is no substitute for the classical value of measurable quantities. For example, for the ground-state hydrogen atom, the electron's expected location is by symmetry in the nucleus. However, since the nucleus is so small as compared with the atom, measurements can never reach it. (Typical calculation would consider a distance equal to the Bohr radius away.) That is good news because if the electron were in the nucleus like a classical particle, its potential energy would be approximately minus infinity rather than the actual value of around −27 eV. It would be farewell to the world as we know it. Despite this, it is, of course, beneficial to have an expectation value than having no value at all.

The "standard deviation" is considered the average difference between the expected value and actual measurements. In the example of the hydrogen atom, where usually the electron is located a distance equal to the Bohr radius away from the nucleus, it turns out that the standard deviation in the x-position is exactly one Bohr radius. The standard deviation is, in general, the objective measure for how much variance a physical value has. If the standard deviation relative to what we are interested in is very high, it is probably safe to use the expected value as a classical value. It is acceptable to assume that the hydrogen atom electron, one is measuring, is in it's place rather than elsewhere. Still, it is not OK for us to believe it has infinite negative potential energy because it is in the nucleus.

3.3.2 Quantum Statistics

Quantum mechanics equations cannot be precisely and fully solved unless a tiny number of particles are present under precise conditions. And then, "exactly" possibly means "exactly in terms of numbers," not analytically. Fortunately, there is good news: Statistical mechanics can make meaningful predictions about large numbers of particles' behavior without writing down each particle's solution. As per the energy spectra, a network of identical fermions (particles with half-integral spin) such as electrons fills the lowest available energy states at absolute zero temperature.

There will be one electron per state (assuming spin is included in the state count, otherwise two electrons per state). The higher energy states will remain unoccupied. In other words, under a dividing energy level, there is one electron per state, and above that energy level, there are zero electrons per state. The energy level, which divides occupied and empty states, is the Fermi energy. For temperatures, T, greater than absolute zero, heat energy makes moving to higher energy levels by at least some electrons. The law of statistical mechanics, which tells us how many these are the so-called "Fermi–Dirac distribution" on average, predicts the average number of fermions per state to be Fermi–Dirac distribution:

$$n = \frac{1}{e^{\left(\frac{(\varepsilon - \mu)}{k_B T}\right)} + 1} \tag{3.23}$$

where ε is the body's energy level, k_B is the Boltzmann constant, and μ is some function of temperature T called the "chemical potential." Let us look at the formula. Firstly, n cannot be more than one because the exponential is more than zero. This restriction is as it should be: There can be a maximum of one electron in a state according to Pauli's exclusion principle, so the average per state n must be one or less. Next, the chemical potential μ equals the Fermi energy at absolute zero.

As a result, energy levels ε above the Fermi energy at absolute zero temperature have the infinite exponential term because the denominator in the exponent approaches zero from positive values; hence the exponential approaches infinite, therefore $n=0$; there are zero particles in those states. On the other hand, the exponential argument for energy states below the Fermi level is minus infinity; hence the exponential is zero, therefore $n=1$; those states have one electron in each state. That is what it ought to be, precisely.

There are not many improvements in the math at a temperature slightly above zero, except for states close to the Fermi energy. Specifically, states just below the Fermi energy will lose some electrons on average to states just above the Fermi energy. In states just below the Fermi level, the electrons are near enough to the empty states to leap at them. The affected energy levels spread around the Fermi field to a range of approximately $k_B T$. In the copper example, the affected range is still tiny compared with the Fermi energy itself, even at average temperatures. In other cases, however, the range affected may be considerable.

Identical bosons satisfy statistics quite differently from fermions. The number of particles outside of the ground state follows the "Bose–Einstein distribution" for bosons:

$$n = \frac{1}{e^{\left(\frac{(\varepsilon - \mu)}{k_B T}\right)} - 1} \tag{3.24}$$

Note that bosons do not obey the principle of exclusion; multiple bosons can be occupying the same state. Also, in Bose–Einstein distributions, the chemical potential μ cannot reach the lowest energy point since the total number n of particles in an energy state cannot be negative. Indeed, when the chemical potential approaches the lowest energy level (something that has been observed to occur at extremely low temperatures), the distribution of Bose–Einstein suggests that the number of particles in the lowest energy state is high, in fact, too high to be adequately represented by the distribution. If this occurs, it is called Bose–Einstein condensation. Many bosons then group in the same state of lowest energy and behave as if they were the same particle altogether. Presumably, this is what happens with liquid helium when it becomes a superfluid. When moving about, a particle "cannot get in its way," and superfluid helium travels around effortlessly without the internal friction that other fluids have.

Both the Fermi–Dirac and Bose–Einstein distributions simplify to the classic "Maxwell–Boltzmann distribution" for high energy levels:

$$n = e^{\frac{(\varepsilon - \mu)}{k_B T}} \tag{3.25}$$

which was developed way before the advent of quantum mechanics.

3.4 QUANTUM FIELD THEORY

In the spontaneous emission process, an excited atomic electron spontaneously releases one (or more) photon(s) and ends up in a lower energy state. This phenomenon arises in the absence of any external stimulus and cannot be described using the atom's typical quantum mechanical model. Moreover, they are interpreted as stationary states by the ideal quantum mechanical model of the atomic energy levels. If an atomic electron occupying an atomic energy level were indeed in a stationary state, no spontaneous emission could ever occur since a stationary state has no time-dependent behavior. So, a quantum theory that can accommodate more than one particle is needed.

Of course, we know how to do this. There is a wrinkle, however. Think again of a photon being emitted by an atomic electron. The system's initial state does have an electron. The system's final state has an electron and a photon.

Additionally, atomic transitions can occur in which more than one photon appears/disappears. The number of particles is fixed in the usual quantum mechanical formalism. Yes, for a particle (or for particles), the normalization of the wave function is seen meaning that the particle (or particles) is (or are) still somewhere. Using traditional quantum mechanical models, we would not be able to explain such processes.

There are also situations where a photon may disappear, producing a pair of electron–positrons and, conversely, where an electron–positron may turn into a photon. However, even the electrons and positrons are not immune to the effects of appearance/disappearance. These variable-particle processes can be understood using the axioms of quantum theory, provided the use of these axioms is smart enough. This improved use of quantum mechanics is usually called quantum field theory, as it can be viewed as an application to continuous systems of the fundamental axioms of quantum mechanics. The picture that emerges is that the building blocks of matter and their interactions are made up of neither particles nor waves but a new form of entity: a quantum field.

Quantum mechanical particles and their duality of the wave particles then become elementary quantum field manifestations. A quantum field describes any elementary particle (although groupings by type depending on the model's sophistication). It has an electron field, a photon field, a neutrino field, and a quark field. In explaining the action of a wide range of atomic and subatomic phenomena, quantum field theory (QFT) has led to remarkable success. The achievement is not merely qualitative; some of the most precise measurements reported include minute atom spectra properties. These properties are predicted through the quantum field theory, and to date, the experiment agreement is outstanding.

The first attempts at quantizing free field theories started only one year after Heisenberg, Born, and Jordan discovered quantum mechanics. Dirac published the first quantum theory of interacting forces one year later, quantum electrodynamics (QED). A year later, Jordan and Pauli introduced the comprehensive system of quantization of relativistic field theories. With the discovery of the gauge field theory and the Standard Model of particle physics' development, the outlook has shifted dramatically. Today, the theory of the quantum field is the basic structure of fundamental physics. QFT provides a unified framework within which the quantum theory and relativity theory are consistently integrated. The QFT revolution lacked a suitable mathematical instrument. The mathematical theory needed to deal rigorously with UV singularities, the theory of the distribution, was developed only in the late 1940s.

The fact that the quantity fields contain distributions is behind UV divergences that require a renormalization system in the quantum field theory. Quantum theory is characterized by a space of states that are vectors $|\psi\rangle$ projective rays of a Hilbert space $\mathcal{H}$. The physical observables in this Hilbert space are Hermitian operators. There is a unique observable in any quantum system, the Hamiltonian $H(t)$, which governs the quantum states' time evolution by a differential equation of the first order. The canonical quantization prescription proceeds by mapping the classical observables, position x, and momentum p into self-adjunct operators $\hat{x}$ and $\hat{p}$ of a Hilbert space $\mathcal{H}$ and replacing the Poisson bracket $\{x:p\}$ by the operator $\left[\hat{x},\hat{p}\right]$.

However, the ladder of quantum states generated by any $\hat{H}$'s values will contain negative states of energy unless the ladder's state vanishes before reaching negative eigenvalues. So, the only possibility consistent with Hamiltonian $\hat{H}$'s positivity is the presence of a final ground state such that $E_0 = 0$. However, the ground-state energy $E_0 = \frac{1}{2}\hbar\omega$. Unlike the classic vacuum configuration energy, E_0 is nonvanishing. In the quantum field theory, the nontrivial value of the quantum vacuum energy has extraordinary implications for vacuum physics.

The most exciting example of field theory is related to relativistic fields. Quantum theories' convergence with relativity theory is not immediate. The first efforts to formulate a quantum phenomenon consistent with the relativity principle leads to confusing theories riddled with paradoxes. The answer to such puzzles emerges from the quantization of the classical field theories. A quantum field theory is a relativistic invariant quantum theory where there is a particular form of quantity operators connected with the classical fields.

In field theory, action S is no more a function of a set number of coordinates but instead of fields. These fields are functions described in an n-dimensional space–time, parameterized by time t and spatial coordinates $n-1$. There are many theories of this kind, but the simplest is based on scalar fields $(\vec{x}, t)$. Here the field is the dynamic variable, and the coordinates $\vec{x}$ should be viewed as labels; they simply define the value ϕ in space and time at a given point. For example, a standard action for a single scalar field $(x)=(\vec{x}, t)$ is provided by:

$$S[\phi] = \int dt \int d^{n-1}x \left\{ \frac{1}{2}(\partial_t \phi)^2 - \frac{1}{2}(\vec{\nabla}\phi)^2 - \frac{1}{2}m^2\phi^2 \right\}$$

$$= \int dt \int d^{n-1}x \ \mathcal{L}(\phi(x), \partial_\mu \phi(x)) = \int L dt \qquad (3.26)$$

which is indeed an invariant of Lorentz, since we have adopted a metric corresponding to the Lorentz-invariant internal product $x^2 = \vec{x}^2 - t^2$. The symbol L is called the Lagrangian density, because its integral over space defines the Lagrangian L. The corresponding Hamiltonian $H[\phi, \pi]$ is given as:

$$H[\phi, \pi] = \int d^{n-1}x \left\{ \frac{1}{2}\pi\dot\phi - \mathcal{L} \right\} = \int d^{n-1}x \left\{ \frac{1}{2}\pi^2 + \frac{1}{2}(\vec{\nabla}\phi)^2 + \frac{1}{2}m^2\phi^2 \right\} \qquad (3.27)$$

Here, $\pi(\vec{x}, t) = \dot\phi(\vec{x}, t)$.

The classical vacuum solution is distinct in this case, $\phi=0$. In the massless case $m=0$, the vacuum degenerates; however, as every stable configuration, ubiquitous ϕ=constant is a solution with finite energy. These configurations have infinite energy in the massive case. As this field theory describes only an infinite set of harmonic oscillators, its quantization is obvious. Like field, this can be decomposed into creation and absorption operators operating on a quantum mechanical Hilbert space. So, we infer that $\left[\phi(x), \dot\phi\left(x' \right) \right]=0$ if a space-like distance separates x and x'. This phenomenon is known as local commutativity, which is a fundamental property that should satisfy any theory of relativistic fields. Local quantity operators take commutes at points not causally related.

3.5 PERTURBATION THEORY

To examine the atomic transitions stimulated by the quantum electromagnetic field, we need specific preliminary results for transitions from the time-dependent perturbation theory, and in particular, "Fermi's Golden

Rule." These are official results of quantum mechanics, and we may take them for granted. Still, we would like to study the derivation of these results because it is vital to precisely see approximations that produce the common lore of atomic transitions.

Let us symbolize the interaction part of the Hamiltonian by the operator $\hat{V} \therefore \hat{H} = \hat{H}_0 + \hat{V}$. One result of time-dependent perturbation is that provided the effects of $\hat{V}$ are small (and assuming the operator $\hat{V}$ is not time-dependent), if the system starts in the initial eigenstate ϕ_i of $\hat{H}_0$ with energy E_i, the likelihood of transition to another eigenstate ϕ_f of $\hat{H}_0$ with energy E_f after the interval of time t is given by the following equation:

$$P(i \rightarrow f) = \frac{4|\phi_f|V|\phi_i|^2}{(E_f - E_i)^2} \sin^2\left\{\frac{(E_f - E_i)t}{2\hbar}\right\}, \ \phi_i \neq e^{i\alpha}\phi_f \qquad (3.28)$$

Provided V's effects are small (and assuming the operator V is not time-dependent), it is believed that the steady, stationary states are part of the discrete spectrum. One can see that the effect of the perturbation described by $\hat{V}$ is to induce transitions with a sinusoidally varying probability given that $E_f \neq E_i$. Note that following the concept of time–energy uncertainty, the timescale Δt for the sinusoidal function to take a considerable value depends on the energy difference $\Delta E = (E_f - E_i)$ according to $\Delta t \Delta E \sim \hbar$. As the final (unperturbed) energy is progressively different from the initial energy, the likelihood of transformation is suppressed by the factor in front of the sine function.

It is possible to have $E_f = E_i$ if the eigenvalue E_i is degenerated. Although these so-called "energy-conserving transitions" (if any) will dominate, the probability of "energy-conserving transitions" is low but nonzero. Here, one should bear in mind that it is just a way of thinking about the transitions in the unperturbed system. The system's actual energy is determined by $\hat{H}$'s eigenvalues and the system's actual energy, and the probability distribution for $\hat{H}$'s eigenvalues is conserved. If the system begins in an eigenstate of $\hat{H}$, it stays in that eigenstate for all times.

The sinusoidal behavior in time of likelihood of transition is an illusion because of a resonance effect for energy-conserving transformations (if any):

$$P(i \rightarrow f) = \frac{|(\phi_f|V|\phi_i)|^2}{\hbar^2} t^2, \ \phi_i \neq e^{i\alpha}\phi_f, \ E_f = E_i \qquad (3.29)$$

One can see that the likelihood of energy-conserving transitions increases squarely over time. Since the perturbation expansion is valid only for sufficiently small transition probabilities, the perturbative formula is only valid in the energy-conserving case for sufficiently short elapsed time, depending on the transition matrix item's size. We have shown that the perturbation effect for transitions between unperturbed stationary states associated with the discrete spectrum of $\hat{H}_0$ is to trigger a periodic transition to and from the initial state, with "energy-conserving" transitions (if any available) becoming dominant after a sufficiently large time interval. Here "large time" means that the time elapsed is much larger than the transition probability oscillation period for energy-non-conserving transitions:

$$t \text{ "} T \equiv \frac{2\pi\hbar}{\left| E_f - E_i \right|} \tag{3.30}$$

Note that the typical energy scale for the atomic structure is in the order of electron volts. This time translates into a standard timescale of $T \sim 10^{-15}$ s, so "large time" is also an excellent approximation for atomic systems in the sense of Equation 3.30, even though the time might not be short enough to use the principle of first-order perturbation. In the above, we assumed that the final state is an energy peculiarity originating from the discrete portion of the energy continuum. In several significant examples, the final state of interest lies (at least roughly) in a continuum. This example will be the case in interest to us because a photon whose wave vector changes continuously (for large enough L) will be present in the final state. In this case, new features appear qualitatively: Periodic transitions are replaced by permanent transitions away from the initial state. We can see that the likelihood of change at "large times" still favors energy-conserving transitions. Again, it will only expand linearly with time because, with time, the width of the distribution of likelihoods about these transitions is growing smaller.

We can see this by assuming that the final state is part of a continuum of states marked by the continuous (unperturbed) energy E and possibly some other observables that are generally denoted by ξ. The probability of transition is now a density of probability (at least in energy space) and must be integrated/summed over a particular range in the variables $(E; \xi)$ to obtain a transition probability. If $t \gg T$, the density of probability can be determined using the representations of the delta function $\delta(x)$, x being a continuous variable:

$$P(i \to f) = \frac{2\pi}{\hbar^2} \left| V_{fi}(E,\zeta) \right|^2 \delta(E - E_i)t, \quad t \gg T \ \& \ V_{fi} \equiv \left\langle \phi_f \middle| \hat{V} \middle| \phi_i \right\rangle \quad (3.31)$$

As can be seen, the transition rate density dP/dt is nonnegligible only for energy-conserving transitions at sufficiently delayed times and with a quasi-continuum of final states. It is constant in time, as opposed to the transitions to discrete spectrum states, a result that is one Fermi Golden Rule version.

In situations where a spectrum of final energies is present, we are interested in the rate of transition to final states with energy in some range E. To obtain a transition rate, we need to integrate the transition rate value over E and sum/integrate over the remaining variables ξ. The result will include a factor of the density of states $\rho(E, \xi)$, which relates an integral over energy to an integral over states. The quantity $\rho(E, \xi)dE$ is the number of states with energies between E and $E + dE$. The density of states will also depend on the discrete variables, even though they are continuous. The rate of transition w from the initial ϕ_i to ϕ_f states with energy E is then expressible in the form:

$$\omega = \sum_\xi \int_E \rho(E,\zeta) dE \frac{2\pi}{\hbar} \left| V_{fi} \right|^2 \delta(E - E_i) = \frac{2\pi}{\hbar} \sum_\zeta \rho(E,\zeta) \left| V_{fi} \right|^2 \quad (3.32)$$

Equation 3.32 is the most used version of Fermi's Golden Rule in the perturbation theory.

CHAPTER 4

Probabilistic Physics

I N PHYSICS, WE SOON realize that the universe is too complicated for us to examine anything in one go. Only if we break it into small parts and research them separately, will we make progress. At the microlevel, quantum mechanics explain elementary particles' behavior, such as the decay of radioactive atoms, the interaction of light with matter, and electrons with magnetic fields, using probabilistic rules as their first principles least in their traditional definition. On the macrolevel, statistical mechanics, in its account of thermodynamic behavior, calls for probabilities in its approach to equilibrium and the second law of thermodynamics. These two entries of probability in modern physics are somewhat distinct.

The philosophical analysis of the foundations of a theory often lags the discovery of the mathematical results that form its basis in the exact sciences' philosophy. The probability theory is no exception. Questions concern both the differences between formalism and the intuitive conceptions of probability and the interrelationships between the intuitive concepts. Often, we will create a mathematical model that reproduces certain features of one of those pieces, and we believe progress has been made. As knowledge progresses, we can develop more and more models, more accurately reproducing features of the real world. No one knows whether this process will end naturally or whether it will continue indefinitely.

Furthermore, since through, definition of the notion of probability is usually intended to be adequate throughout, regardless of context, the various applications of probability theory pull in different interpretative directions: Some applications, say in decision theory, are likely to be subjectively interpreted as representing the degree of belief of an individual,

DOI: 10.1201/9781003206743-4

while others, say in genetics, call for an objective notion of probability, which characterizes certain biological phenomena.

Probabilistic laws are taken in modern quantum mechanics to replace traditional mechanics and classical electrodynamics. Quantum probabilities are here interpreted as representing real stochastic behavior, ungoverned by deterministic rules. On the other hand, in classical statistical mechanics, probabilistic laws are added to, and perhaps even reducible to, the underlying deterministic mechanics. Consequently, these uncertainties are viewed as reflecting our ignorance of the existence of microstates in the world or as a by-product of our gross-grained descriptions of these states.

Despite this radical distinction, both theories of probabilistic laws pertain to the behavior of actual physical systems. When quantum mechanics attribute the same probability to a radium atom's decay, it must be suggesting something about the atom, not just our beliefs, expectations, or knowledge of the atom. Moreover, when classical statistical mechanics attribute a high likelihood of disseminating a gas across the volume that is available to it, it is supposed to say something about the gas, not just about our subjective beliefs. The probabilistic laws in both quantum and statistical mechanics should have some legitimate objective substance in this context. However, this factual content and its relation to epistemic notions such as ignorance, rational belief, and descriptions' accuracy are open issues.

Since statistical mechanics was developed, probabilistic analysis played an increasingly significant role in physics. During quantum mechanics formation, the latest critical phase in developing the statistical approach to physics was taken. The pioneers of quantum theory understood that the definition of physical processes for individual elementary particles could not be provided by quantum formalism. Understanding this startling finding prompted various discussions about the possibilities of personal and probabilistic representations and their relationships. A wide variety of opinions on the roots of quantum stochasticity characterize these debates.

One of the perspectives is the disparity between "quantum stochasticity" and "classical stochasticity." Quantum mechanics, therefore, could not be reduced to traditional statistical mechanics. This viewpoint suggests a traditional understanding of quantum mechanics. We do not use empirical realism in the quantum definition of truth from this understanding. The most basic physical quantities, such as the position and momentum of an elementary particle, could not be regarded as the entity's properties, the particle. The elementary particle can be in a state where alternatives are superimposed. Only the act of a calculation gives these alternatives choices of the probabilities.

Besides, the entire "quantum house" was based on two experimental cornerstones: (1) the photoelectric emission experiment and (2) the two-slit experiment. The first experiment indeed revealed that light occupies the corpuscular structure (discrete energy structure). The second experiment, however, revealed that photons do not obey traditional classical statistics. The conventional rule for introducing probabilistic alternatives is $P = P_1 + P_2$ in the interference experiment; it is broken. Instead of this law, probabilities observed in interference experiments obey quantum law for probabilistic alternatives to be added: $P = P_1 + P_2 + 2\sqrt{P_1 P_2} \cos\theta$. So, the classical law is usually disrupted by the $\cos\theta$ factor.

Again, we emphasize that all these radical changes had a strictly probabilistic origin, namely the introduction of the new probabilistic law. We also emphasize that the founders of quantum mechanics did not have a profound probabilistic analysis of the issue. Instead, they were studying other physical model components. Such an approach induces the current physical reality concept we have already discussed, namely "quantum reality."

4.1 PROBABILITY THEORY

Assume a policeman strolls down a deserted street on a dark night. Unexpectedly, he hears a security alarm, looks across the street, and sees a smashed window in a jewelry store. A young man wearing a mask then emerges out, crawling through the broken window, carrying a bag that turns out to be full of expensive jewelry. The officer does not hesitate at all to determine this person is unethical. Nevertheless, what method of deduction does he use to come to this conclusion? Let us look at the general essence of these issues first. A reflection of the moment clarifies that our policeman's inference was not a rational deduction from the evidence, for there might have been a completely innocent reason behind it all. For example, it may be that this man was the jewelry shop owner, and he came home from a fancy-dress party and did not have the key with him. Just as he was walking by his store, a passing truck threw a stone through the window, and he only protected his own property.

Nevertheless, while the policeman's rational framework was not a reasonable inference, we must admit that it had a degree of validity. The data did not make the dishonesty of the gentlemen definite, but it made it highly probable. This scenario is an example of a sort of logic in which we all were skilled, naturally, long before we learned mathematical theories. We will scarcely go through a waking hour without facing a situation (e.g. will it rain or won't it?) in which we do not have enough knowledge to encourage deductive reasoning, but we do have to know what to do instantly.

Despite its familiarity, the forming of rational assumptions is a very subtle process. While history documents discussions of it stretching over centuries, certainly no one has ever created an interpretation of the method that anyone else considers entirely satisfactory. The problem is resolved by some excellent and motivating modern developments in which definite theorems replace contradictory intuitive assumptions, and ad hoc procedures are replaced by principles that are uniquely defined by some relatively simple – and almost unavoidable – reasoning standards. These modern developments also mark the emergence of probability theory.

Probability is the science of nondeterministic or random experiments. If a dice is thrown in the air, then the die will surely come down, but it is not certain that, say, a 6 will appear. Nevertheless, suppose we repeat this experiment of shooting a die; let s be the number of successes, i.e. the number of times a 6 appears, and let n be the number of shootings. It was then empirically found that the ratio $f = s/n$, called relative frequency, is constant in the long run, i.e. approaches a limit. The foundation of probability theory is this stability. The probability p of an event A is described as follows: if A can occur in s ways out of an equally probable total of n ways, then $p = P(A) = s/n$.

For example, an even number can occur in three ways out of six "equally probable" ways in tossing a die; hence, $p = 3/6 = 1/2$. This classical concept of probability is circular because the idea of "equally likely" is the same as that of undefined "with equal probability." If we throw our die, indeed, the number that will come up is one of 1, 2, 3, 4, 5, or 6. Thus by the rule just given, unity must be the sum of the probabilities associated with each of these numbers that are coming up. If we know the die is fair, we believe that no number will come up more often than another, so all six odds must be equal. They must, therefore, all be equal to 1/6. We have the rules to generalize this example. With only n mutually exclusive results, $\sum_{i=1}^{n} p_i = 1$.

If all results are equally likely, then $p_i = \dfrac{1}{n}$.

4.1.1 Events and Sample Space

We conduct experiments and observations of the natural world to learn about the state of nature and ask questions about the effects. The topic of questions that we may ask about the results of an experiment is called

events, such that the potential responses are whether it occurs or it does not occur. There are different kinds of events, and we have **the elementary events** among them; that is, those random experiment results that cannot be decomposed in other smaller entities. The **sample space** (like Hilbert space H defined in quantum mechanics) is the set of all possible elementary outcomes (events) of a random experiment and must be the following:

i. Comprehensive: Any potential outcome of the experiment must be included in the sample space.

ii. Exclusive: There is no overlap of elementary results.

To research random phenomena, we begin by defining the sample space, so we need to understand the experiment's potential outcomes. Consider the simple experiment of rolling a die with six faces enumerated from 1 to 6. We consider event e_i = {get the number i on the upper face} as elementary events; i = {1,2,...,6} such that the sample space = {$e_1, e_2, ..., e_6$}. Note that every potential roll outcome is included in the sample space, so we cannot simultaneously have two or more elementary outcomes. However, besides the elementary ones, there are other types of events.

For example, we may be interested in the parity of the number, so we would also like to consider the possible results: A = {die produces an even number} or B = {die produces an odd number}.The result A = {e_2, e_4, e_6} and B = {e_1, e_3 or e_5} are not elementary. In general, an event is any subset of sample space. We can differentiate between elementary events that are an element of sample space and any subset of sample space and events at two extremes. Thus, certain events are any result contained in sample space, and impossible events are any result not contained in sample space. Any event that is neither likely nor impossible is considered a random event. Going back to the dice rolling, likely events are such that {get a number n| $0 \leq n \leq 1$} or {get a number that is even or odd}, and unlikely events are such that{get an odd number that is not prime}or {get a number 7 on dice with six sides}.

Empirically, events can be integrated to create hybrid events: If A is the event that a certain blue die falls with 2 facing up, and B is the event that a red die falls with 6 up, then AB is the event that the blue die shows 2 if both dice are thrown, and the red one displays 6. If A's probability is p_A and the likelihood of B is p_B, then the blue die will show 2 in a fraction of a p_A of the two dice throws, and in a $_{pB}$ fraction of those throws, the red die will have 6 up. Therefore, the proportion of throws in which AB occurs is $p_A p_B$, so we can take AB's probability of being $P_{AB} = p_A p_B$. In this example, A and B are independent events as we see no reason why the number displayed by the blue die should be affected by the number that shows up on the red one and vice versa. Multiplying them is the procedure to combine the probabilities of separate events to obtain the likelihood of all events occurring: p (A and B) $= p(A)p(B)$ (for independent events). Event A above excludes the event C where the red die shows 4; A and C are exclusive events because only one number can come upon a die in each throw. The chance of seeing either a 2 or a 4 is obtained by adding p_A and p_C. Accordingly, $p(A$ or $C) =$ $p(A) + p(C)$ (for exclusive situations).

4.1.2 Expected Value

A random variable x is a measurable number, and the value we get is subject to uncertainty. Suppose for simplicity that only xi can be measured with discrete values. For example, x might be the number that comes up in the case of a die, so x has six possible values, $x_1 = 1$ to $x_6 = 6$. If p_i is the probability that we will measure x_i, then the expected value of x is $x = \sum_i p_i x_i$.

If the event is reproducible, it is simple to show that, as n becomes very high, the average values we calculate on n tests lead to x. Consequently, the average of x is often referred to as x.

Suppose we have two variables at random: x and y. Let p_i be the probability our measurement will return x_i for x and y_j for y. Then the sum expected $x + y$ is:

$$\langle x+y \rangle = \sum_{ij} p_{ij}\left(x_i + y_j\right) = \sum_{ij} p_{ij}x_i + \sum_{ij} p_{ij}y_j \tag{4.1}$$

However, $\sum_j p_{ij}$ is the likelihood we will measure x_i regardless of what we measure for y, so it must be p_i equal. Similarly, $\sum_i p_{ij} = p_j$, the probability of calculating y_j whatever we get for x. Substituting those expressions in Equation (4.1), we get:

$$\langle x+y \rangle = \langle x \rangle + \langle y \rangle \tag{4.2}$$

In other words, the expected value of the sum of the two random variables is the sum of the actual expected values of the individual variables, regardless of whether the variables are independent. Specifically, the expected value for the sum of two random variables is the sum of the individual expected values of the variables, regardless of whether the variables are independent. A useful measure of the degree by which a random variable's value fluctuates between trial and trial is the variance in x, which is defined as

$$\Delta_x^2 = \left\langle \left(x - \langle x \rangle \right)^2 \right\rangle = \left\langle x^2 \right\rangle - \left\langle x^2 \right\rangle \tag{4.3}$$

4.1.3 Probability Amplitude

Many areas of the psychological, physical, and medical sciences make substantial use of probabilities. However, quantum mechanics stands alone in the way it measures probabilities, for it always calculates a probability p as the modulus square of a complex number A: $p = A^* A = |A|^2$. The complex number "A" is called the amplitude of probability p, and A^* represents the complex conjugate. Quantum mechanics is the only branch of science in which amplitudes of probability occur, and no one knows why they come up. By following the basic theory, they give rise to phenomena that have no analogues in classical physics.

Assume something can occur through two (mutually exclusive) paths, S or T, and let the amplitude of probability for it to happen through route S be $A(S)$ and the amplitude of probability for it to happen on route T be $A(T)$. The probability that an incident will occur is not merely the sum of the probability that one of the two alternative routes will occur: There is an additional parameter. There are two ways to combine amplitudes from the two routes to find probabilities for observing combined events. One adds probabilities when the final states can be distinguished: $p_{dis}(S \text{ or } T) = p(S) + p(T)$. If the final state is indistinguishable, one adds amplitudes:

$$p(S \text{ or } T) = |A(S \text{ or } T)|^2$$

$$= A^*(S)A(S) + A^*(S)A(T) + A(S)A^*(T) + A^*(T)A(T)$$

$$= |A(S)|^2 + |A(T)|^2 + 2RA(S)A^*(T) \tag{4.4}$$

In regular probability theory, this concept has no equivalent, and it contradicts the basic principles of probability theory. It relies on the probability amplitude phases for individual paths and does not contribute to the probabilities $p(S) = A^*(S)A(S) = |A(S)|^2$ of the paths. The terms that mix the amplitudes are the "interference" terms. The terms of interference are why we cannot ignore the complex nature of the amplitudes, and they cause many types of quantum phenomenon. The breaches of probability theory caused by quantum interference are not observable in some situations, so standard probability theory applies. How do we then know that the theory in Equation 4.4 is valid and has such extraordinary consequences? The best-informed response is that it is a fundamental postulate of quantum mechanics. If we look at a digital watch, touch a keyboard, listen to a CD player, or communicate with some other electronic device that has been designed with the assistance of quantum mechanics, we are evaluating and validating the theory. Our human civilization now depends entirely only on the validity of Equation (4.4).

4.1.4 Two-Slit Quantum Interference

Let us consider an experiment to clarify the principle's physical implications and propose how it can be tested. The system comprises an electron gun, S, a screen with two narrow slits 1 and 2, and a photographic surface that blackens when an electron hit (see Figure 4.1).

An emitted electron has an amplitude to go through slit 1 and then hit the screen at point x. This amplitude will depend clearly on point x, so we will call it $A_1(x)$. Similarly, the amplitude $A_2(x)$ exists that the electron

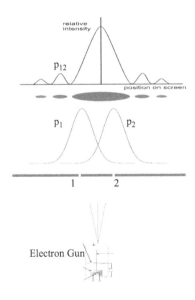

FIGURE 4.1   Double-slit interference experiment with electrons.

passed through slit 2 before hitting the photographic plate on x. Therefore, the probability of the electron arriving at x is:

$$p(x)= \left|A_1(x)+A_2(x)\right|^2 = \left|A_1(x)\right|^2 + \left|A_2(x)\right|^2 + 2RA_1(x)A_2^*(x) \quad (4.5)$$

The probability amplitude $\left|A_1(x)\right|^2$ is simply $p_1(x)$, which is approximately a Gaussian distribution centered on the value x_1 of x. A straight line from the electron gun through the middle of slit 1 hits the plate. Similarly, $p_2(x)$ should be a Gaussian function centered at the x_2 intersection of the screen and a straight line from the gun through slit 2. Both generally have a value as $A_i = \left|A_i\right|e^{i\phi_i}$, where ϕ_i is the phase of the complex number A_i. Then equation can be written as $p(x) = p_1(x) + p_2(x) + I(x)$, where the interference term I is given as:

$$I(x)=2\sqrt{p_1(x)p_2(x)}\cos\left(\phi_1(x)-\phi_2(x)\right) \quad (4.6)$$

Let us observe the $I(x)$ behavior near the equidistant points from the slits. Then $p_1 \approx p_2$ (see Figure 4.1), and the interference term is comparable in magnitude to p_1+p_2, and, by Equation 4.6, the probability of an electron arriving at x will oscillate between $\sim 2p_1$ and 0 depending on the value of the phase difference $\phi_1(x) - \phi_2(x)$. Let us substitute machine gun bullets for the electrons. Then ordinary experience tells us that classical physics applies and indicates that the probability $p(x)$ of a bullet arriving at x is just the sum $p_1(x) + p_2(x)$ of the probabilities of a bullet coming through slit 1 or 2. Classical physics, therefore, does not predict a sinusoidal pattern in $p(x)$. How do we reconcile the predictions of classical and quantum mechanics? Why is the compromise that bullets do not follow quantum mechanics? We believe they do, and there should be a sinusoidal pattern in the probability distribution for the arrival of the bullets. However, we find that quantum mechanics predicts that the distance Δ between this pattern's crests and troughs becomes narrower and narrower as we increase the mass of the particles we are firing through the slits. When the particles are as massive as a bullet, Δ is phenomenally small as 10^{-29} m. Consequently, testing whether $p(x)$ becomes small at regular intervals is not experimentally feasible. Any feasible experiment will prove the average $p(x)$ value of the sinusoidal pattern over many peaks and troughs, and the average $I(x)$ value in Equation 4.6 vanishes.

4.2 PROBABILITY DISTRIBUTIONS

Before we discuss the probability distributions, we need to define a few entities. When we experiment, we observe an outcome in the form of the data,
and these data could be of many types, as described in Chapter 1. For con-
venience and general use, the data can be represented by an entity called
random variable; let us say x. It is termed a random variable if we have
a variable, and we can find a probability associated with that variable. In
many cases, the random variable is what we measure, but it is usually what
we count is ascribed as discrete random variables. The random variable is
the person's height in each classroom that could be measured on a continuous scale. For the same example, if we are counting the number of students
in a class, the random variable is the number of students in the class.

Now assume we put all the random variable values together with the probability that random variables will occur. We would then have a distribution,
except now it is called a probability distribution since it includes probabilities.
A probability distribution is a probability assignment of random variable
values. The pdf abbreviation is used for a probability distribution function.
For probability distribution function:

$$0 \le P(x) \le 1 \text{ and } \sum P(x) = 1 \tag{4.7}$$

We must always ensure that the specified values, intervals, or categories are
mutually exclusive and generally comprehensive when defining the results
of a random process. We shall ensure that the list contains all possible
results and that the results do not overlap. It is often possible but can also
need some consideration. Provided this requirement is met, this result's
likelihood is merely equal to the sum of their probabilities.

One can calculate the mean and standard deviation in the same way as
with any data set. In problems involving a probability distribution function (pdf), one finds the probability distribution of the population often by
experimenting several times. This experimenting is because the data from
repeated experiments are used to estimate the actual probability. Since a
pdf is a population, the calculated mean and standard deviation are the
population parameters and not the sample statistics. The notation used
is identical to the notation used for the population mean and population
standard deviation. One can think of the mean as the expected value. If
the trials were repeated an infinite number of times, it is the value one
expects to get. The mean or predicted value need not be a whole number,

even though the potential x values are whole numbers (as in the number of students in each classroom).

In real-life applications, specific probability distributions occur with such frequency that they had been awarded their names. In the following, we are surveying and studying some of their basic properties. The following distributions will be discussed: binomial, Poisson, normal, uniform, and exponential. The first two are discrete, and the final three are continuous.

4.2.1 Binomial Distribution

There are a set of principles that would result in a binomial distribution, if true, which are as follows:

- There are a range of n trials or experiments.

- Every trial may lead to either a failure or a success (only two possible outcomes).

- For all trials, the probability p of success is the same.

- The findings are independent of the specific trials.

- The total number of successes in those n trials is of interest to us.

Under the assumptions set out above, let X be the total number of successes. Therefore, X is considered a binomial random variable, and X is considered the binomial probability distribution. Let X be a binomial random variable. Then, its probability mass function is:

$$P(X=x)=\frac{n!}{x!(n-x)!}p^x(1-p)^{n-x} \text{ for } x=0,1,2...,n \qquad (4.8)$$

The results of n and p are named distribution parameters. Note that the probability of observing any sequence of n independent trials containing x successes and $n - x$ failures is $p^n(1-p)^{n-x}$. The overall number of possible combinations if we randomly select x objects out of n is:

$$\left(\begin{array}{c} n \\ k \end{array} \right) = \frac{n!}{x!(n-x)!}$$

For binomial distribution, the mean $\mu = E(X)=np$ and variance:

$$\sigma^2 = V(X) = np(1-p) \tag{4.9}$$

For example, we consider that a batch of 1,000 chips is a sequence of $n = 1,000$ Bernoulli tests. Each chip has a probability $p = 0.01$ of being defective. We will be able to use the binomial distribution to answer the probability that no more than one chip out of the 1,000 chips will be defective. Here, we have $x = 1$, and substituting these numbers in Equation 4.8, we get:

$$P(X=1) = \frac{1000!}{1!(1000-1)!} 0.01^1 (1-0.01)^{1000-1} = 0.00044$$

4.2.2 Poisson Distribution

When a set of canonical conditions outlined in the following are sufficiently met, the Poisson distribution emerges.

- The number of occurrences at any time interval is independent of the number of occurrences at any other disjoint interval. Here, the typical definition for an "exposure variable" is "time interval," although other definitions are possible. Example: rate of error per page in a document.

- For all intervals of the same size, the distribution of several events in an interval is the same.

- For a "low" time interval, the probability of observing an occurrence is proportional to the interval length. The constant proportionality corresponds to that of "level" of events occurring.

- The probability of witnessing two or more occurrences in an interval approaches zero as the interval is getting smaller.

Under the assumptions, let λ be the rate at which events occur, t be the length of a time interval, and X be the total number of events in that time interval. Then, X is called a Poisson random variable, and the distribution of X is called the Poisson probability distribution. Let $\mu \equiv \lambda t$; μ can then be interpreted as the average, or mean, number of events in at the interval. The Poisson probability mass function is then:

$$P(X=x) = \frac{\mu^x}{x!} e^{-\mu} \text{ for } x = 0,1,2,... \tag{4.10}$$

The μ value is the distribution parameter. In any specific time, μ may be defined as λ times the length of the interval for a given time interval of interest. For Poisson distribution, the mean $E(X) = \mu$ and variance $V(X) = \mu$. For example, consider that one nanogram of Plutonium-239 will have an average of 2.3 radioactive decays per second, and a Poisson distribution approximates the number of decays. What is the likelihood that there will be exactly three radioactive decays in two seconds? X = # of decays per time interval of interest (two seconds), λ = 2.3 per second $x2 = \mu = 4.6$ per two second interval, x = # of decays to be exactly 3, the probability is given as:

$$P(X=3) = \frac{4.6^3}{3!} e^{-4.6} = 0.163.$$

4.2.3 Normal Distribution

The most important distribution is normal distribution, as it occurs naturally in various applications. The main explanation for this is that a large number of random variables always turn out to be distributed normally; a random variable X is assumed to have the normal distribution with parameters μ and σ if its density function is given by:

$$f(x) = \frac{1}{\sigma\sqrt{2\pi}} e^{-\frac{1}{2}\left(\frac{x-\mu}{\sigma}\right)^2} \quad for -\infty < x < \infty \tag{4.11}$$

The mean is given by the expected value $E(X) = \mu$ and the variance $V(X) = \sigma$. Thus, a mean μ and a standard deviation σ define the normal distribution. The curve is bell-shaped that is symmetric around the mean μ. The standard deviation σ controls the curve's "flatness." Increasing the mean moves the curve of density to the right, and increasing the normal deviation will flatten it, as shown in Figure 4.2.

A normal distribution with a mean of 0 and a standard deviation of 1 is called the standard normal distribution. For this case, the probability density function takes the simpler form:

$$f(x) = \frac{1}{\sqrt{2\pi}} e^{-\frac{x^2}{2}} \quad for -\infty < x < \infty \tag{4.12}$$

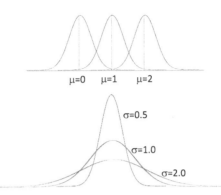

FIGURE 4.2   Normal distribution with different means and standard deviations.

A common practice is to convert a standard random variable X with arbitrary parameters μ and σ through the transformation into a standardized standard random variable Z with parameters 0 and 1:

$$Z = \frac{X - \mu}{\sigma} \tag{4.13}$$

For example, consider a box containing 100 Ω resistors reported to have a standard deviation of 2. What is the probability of choosing a 95 Ω or less resistor? What is the probability of having a resistor in the 99–101 Ω range? We are given $\mu = 100$ and $\sigma = 2$, so from Equation 4.13, we get:

$$Z = \frac{X - 100}{2},$$

for 95 Ω, $Z = (95-100)/2 = -2.5$, using standard z-tables or Microsoft Excel, we have to find $P(Z \le -2.5) = 0.0062$, and for the probability of finding a resistor in the range 99 and 101, we find $P(-0.5 \le Z \le 0.5) = 0.3829$.

4.2.4 Uniform Distribution

The simplest example of a continuous probability distribution is the uniform distribution. If a random variable X is said to be distributed uniformly, its density function is given by:

$$f(x) = \frac{1}{b-a} \ for -\infty < a \le x \le b < \infty \tag{4.14}$$

An example of a uniform distribution is shown in Figure 4.3.

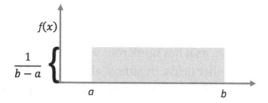

FIGURE 4.3   Visual of a uniform distribution.

where the shaded region has an area (width times height) $(b - a)$ $[1/(b - a)] = 1$. For example, consider that a heating and air conditioning service company reports that the amount of time to fix a furnace is spread evenly between 1.5 and 4 hours. Let x = time it took to fix a furnace. Find the probability that it takes more than 2 hours to repair a randomly selected furnace. Here, $b = 4$ and $a = 1.5$ so from Equation 4.14, $f(x) = 0.4$ and $P(x > 2) = (base)(height) = (4 - 2)/(0.4) = 0.8$.

4.2.5 Exponential Distribution

The exponential distribution, which has the following probability density function, is another useful continuous distribution:

$$f(x) = e^{-x} \ for \ x \geq 0 \tag{4.15}$$

A single parameter λ, which is called the rate, characterizes this family of distributions. Intuitively, λ can be regarded at any time t as the instantaneous "failure rate" of a "device," since the device has survived up to t. Usually, the exponential distribution is used to model time intervals between "random events." If a random variable X is distributed exponentially at the rate λ, the mean is given by the expected value $E(X) = \dfrac{1}{\lambda}$ and variance is $V(X) = \left(\dfrac{1}{\lambda}\right)^2$. For example, an alkaline battery's lifetime X is distributed exponentially, with $\lambda = 0.05$ per hour. What is the mean and standard deviation for the battery's lifetime? Response: $E(X) = SD(X) = 1/0.05 = 20$ hours. What are the probabilities that the battery will last 10–15 hours and last longer than 20 hours? Reply:

$$P(10 < X < 15) = e^{-0.5*10} - e^{-0.5*15} = 0.1341 \text{ and } P(X > 20) = e^{-0.5*20} = 0.3679.$$

4.3 CENTRAL LIMIT THEOREM

The central limit theorem is a powerful tool in statistical inference and mathematics in general, as it has numerous applications in science, manufacturing, and many other fields. In almost all of the problems, the normal or standard sampling distribution remains much simpler to use and to deliver better results (more reliable parameter estimates) than any alternative sampling distribution that anyone might suggest.

Typically, one will use the maximum likelihood tool in estimating the mean μ for data D; that is, the value of μ, for which the probability of finding μ is a maximum. If the preceding information is unimportant (i.e. if the initial probability density is nearly constant over the high-probability region), the Bayesian might do that as well. Nevertheless, this method of the maximum likelihood yields the most accurate estimates? Will the estimates of μ not be made even better by a different option in the long run (i.e. more closely based on the actual value μ_0)? This problem is precisely what the central limit theorem answers.

A random sample of size n of a given distribution is a set of X_1, X_2,. X_n independent random variables, X_n, each with a given distribution, with $E(X_i) = \mu$ expected value and $V(X_i) = \sigma^2$ variance. Such a set of variables is often called random sampling distribution and has the following properties:

- Sample sum: $S = \sum_{i=1}^{n} X_i$, $E(S) = n\mu$ and $V(S) = n\sigma^2$

- The sample mean: $\bar{X} = \dfrac{1}{n}\sum_{i=1}^{n} X_i$, $E(\bar{X}) = \mu$, and $V(\bar{X}) = \dfrac{\sigma^2}{n}$

The central limit theorem is concerned with drawing finite samples of size n from a population with a known mean, μ, and a known standard deviation, σ. The first alternative of the theorem is to say that if we collect samples of size n and n as "large enough," calculate each sample's mean and create a histogram of all those means. The resulting histogram tends to have an approximate bell-shaped normal distribution. The second alternative says that if we recollect samples of size n that are "large enough," measure each sample's total, and create a histogram, then the resulting histogram will again appear to have a standard bell-shaped normal distribution. In either case, shape of the original population distribution does not matter at all.

The large numbers law says that if one takes more extensive and larger samples from any population, the sample's mean $\bar{x}$ will be closer and closer to μ. We know that the sample averages follow a normal distribution from the central limit theorem, as n becomes larger and larger. The greater n becomes, the smaller becomes the standard deviation. (Remember, the standard deviation for X is $\sigma/\sqrt{n}$.) This rationale assumes that the mean $\bar{x}$ of the sample would be identical to the mean μ for the population. We may say μ is the value that the sample averages approach as n gets increased. Therefore, samples can reflect population attributes, such as mean and variance, with some degree of confidence.

4.3.1 Confidence Level and Interval

An essential role of the central limit theorem is that it allows us to make statements about the population from which it was chosen using information collected from a sample. How good is that sample providing us with an estimate? For example, there are various ways to estimate. Using the expected value as an estimation of the mean population is known as the estimated point. An alternate is an interval estimate that provides a range of values for the population parameters. It helps us deal with uncertainty in obtaining the population parameters and provides a range that may include the population parameters with a certain level of confidence. The interval that is calculated to contain the population parameter is known as the confidence interval. The level of certainty with which we can say that it could contain the population parameter is known as the confidence level.

Let us consider a known variance σ^2 for an arbitrary random variable X, and let $X_1, X_2, ..., X_n$ be a large size n random sample from the X distribution. And let $Z = N(0, 1)$ be a standard normal variable, and let $F_Z(-z_{\alpha/2}) = \alpha/2$ be such that $z_{\alpha/2} > 0$. Then, the random interval $[L, R]$, where

$$L = \bar{X} - z_{\alpha/2}\frac{\sigma}{\sqrt{n}} \quad \text{and} \quad R = \bar{X} + z_{\alpha/2}\frac{\sigma}{\sqrt{n}} \quad \text{is an interval of confidence}$$

for the actual population mean μ with confidence level $1 - \alpha$, i.e. $P(L \leq \mu \leq R) = 1 - \alpha$.

For example, consider all hypertensive males who smoke and are interested in the distribution of serum cholesterol levels for them. This distribution has a standard deviation of 46 mg/100 mL and an unknown mean μ. Suppose we draw a random sample of 12 individuals from this population and find the mean cholesterol level is $\bar{x} = 217$ mg/100 mL. This calculation is a point in time estimate of unknown mean population cholesterol level μ.

However, it is important to create an interval estimate of μ to account for the sampling variability. A confidence interval of 95% for μ is:

$$\left(217 - 1.96\frac{46}{\sqrt{12}}, 217 + 1.96\frac{46}{\sqrt{12}}\right) \text{ or } (191, 243). \text{ A confidence interval of}$$

99% for μ is:

$$\left(217 - 2.58\frac{46}{\sqrt{12}}, 217 + 2.58\frac{46}{\sqrt{12}}\right), \text{ or } (183, 251).$$

4.4 HYPOTHESIS TESTING

Hypothesis testing is an attempt to construct a framework of justification in mathematics. Our minds are influenced by evidence rather than by arguments. Therefore, to determine how our mind is "guided in one way or the other" by new evidence, we investigate some applications which, while mathematically simple, have proven to be of functional significance in many distinct fields. The hypothesis testing approach uses sense tests to assess the possibility that a claim (often linked to the mean or variance of a given distribution) is valid. With that probability, we will accept the statement as valid. While it is essential to understand the mathematical concepts that go into the formulation of these tests, it is even more important to know how to appropriately use each test (and when to use which test).

We start by formulating the hypothesis we would like to test, called the alternative hypothesis. This hypothesis is usually derived from an attempt to prove a theory that underlies it. We do so by testing the alternative hypothesis against the null hypothesis. Finally, we set a level of probability α; this value will be our level of significance and corresponds to the likelihood of rejecting the null hypothesis when it is true. The rationale is to conclude that the null hypothesis is valid and then perform an analysis on the parameters concerned. Suppose the analysis shows unexpected outcomes for the null hypothesis to be valid (like outcomes that can only occur with a minuscule probability, say 0.01). In that case, we can confidently assume that the null hypothesis is not accurate and support the alternative hypothesis. Now that we have defined the hypotheses and the level of significance, and the data are collected. Once the data have been compiled, hypothesis tests can proceed as per the type of test we need to perform and test statistics that need to be used.

4.4.1 Types of Error in Hypothesis Testing

As explained above, since we observe a sample and use the sample parameters to arrive at a decision we make instead of the whole population, a conclusion may be wrong. Table 4.1 shows that there are four alternatives to the decision we make about a null hypothesis regarding the truth and the falsity, and these are also represented graphically in Figure 4.4:

1. The decision on maintaining the null hypothesis may be correct.

2. The decision to keep the null hypothesis could be wrong.

3. The decision to reject the null hypothesis can be correct.

4. The decision to reject the null hypothesis may be incorrect.

If we decide to reject the null hypothesis, then we may be right or wrong. The worst wrong decision would be to dismiss a real null hypothesis. The decision is an example of a type I error. With every test we do, there is always a distinct possibility that our decision is a type I error. A researcher making this mistake decides to reject preceding notions of truth that are actually true. Making that kind of mistake is analogous to finding an innocent person guilty. To mitigate this mistake, we conclude that a defendant is innocent at the outset of a trial. Similarly, we assume that the null hypothesis is correct when starting a hypothesis test to minimize making a type I error or an α-error.

Similarly, if we decide to retain the null hypothesis, then we may be right or wrong. The right decision is to withhold a correct null hypothesis. This decision is termed a null finding. Usually, this is an uninteresting decision as the decision is to retain what we have already assumed: that the value stated in the null hypothesis is correct. Because of this, null results alone are rarely published in research.

TABLE 4.1   Showing All the Possible Scenarios and Types of Errors That Can Be Committed in Decision-Making

		Actual situation	
		H_0 is true	H_0 is false (H_1 is valid)
Decision based on the test statistics	Reject H_0 and accept H_1	Type I error (α)	Correct decision $(1 - \beta)$
	Fail to reject H_0	Correct decision $(1 - \alpha)$	Type II error (β)

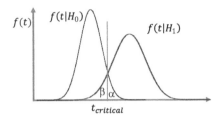

FIGURE 4.4 Visual meaning on hypothesis testing and type I and type II errors.

The wrong decision is to hold onto a false null hypothesis. This decision is an example of a type II error or β-error. There is also a chance with every test we do that the decision may be a type II error. We decide in this decision to retain preceding notions of truth that are false. Furthermore, if it is a mistake, we have not done anything; we kept the null hypothesis. We can still go back and do more research and eventually find a confidence level to reject the null hypothesis. The power in the testing hypothesis is the likelihood of rejecting a false null hypothesis. Specifically, when the null hypothesis is indeed false, it is the probability that a randomly selected sample will show that the null hypothesis is false. Mathematically, it is equal to $1 - \beta$.

4.4.2 Types of Tests and Test Statistics

The test statistics transform the sampling distribution we observe into a normal standard distribution, thus allowing us to decide. The test statistics we use are mostly based on what we know about the population. We can use the z test if we know that the population is normally distributed or the sample size $n \geq 30$, or a t-test if the sample size $n < 30$. We can use the one independent sample t or z test when we know the mean and standard deviation within a single population that is normally distributed. The one independent sample t or z test is a statistical technique used to test hypotheses about the mean of a known variance in a single population.

In a nondirectional or two-tailed test, where the alternative hypothesis is reported as not equal to the null hypothesis, we will place the significance level in both tails of this test's sampling distribution. Therefore, from the null hypothesis, we are interested in some alternatives. It is the most popular alternative theory that has been tested in scientific research. Nondirectional tests, or two-tailed tests, are tests of hypothesis where the alternative hypothesis is specified as not equal to ($\neq$). Any alternative from the null hypothesis is of interest to the researcher.

One of three alternative hypotheses can be stated: a population mean is greater than (>), less than (<), or not equal to (≠) the value of a null hypothesis.

4.4.2.1 z-test

The first form of test that we are investigating is the most basic: measuring the mean of a distribution in which we are already familiar with the variance in population σ^2. Later, we will explore how to change these assessments to deal with the case where we do not know the population's variation. Thus, we conclude that our population is natural for the time being natural with a known variance of σ^2. Our test statistics are $z = \dfrac{\bar{x} - \mu}{\sigma/\sqrt{n}}$ where n is the number of observations made when the study data are obtained, and μ is the real mean when we conclude that the null hypothesis is correct. Furthermore, to test a hypothesis with a given significance level α, we calculate the critical value of z (or critical values, if the test is two-tailed) and then check to see whether the test statistic value is in our critical region or not. It is called a z-test. Tests involving either $\alpha = 0.05$ or $\alpha = 0.01$ are most frequently used. As we construct our crucial region, we have to determine whether our hypotheses are one-tailed or two-tailed. When one-tailed, the null hypothesis would be dismissed when $z \geq z_\alpha$ or if $z \leq z_\alpha$. If two-tailed, the null hypothesis would be dismissed if be $|z| \geq z_{\alpha/2}$.

4.4.2.2 t-test

Unfortunately, z-tests need one of two conditions: either the population with known variance is normally distributed, or the sample size is large. The underlying distribution is roughly normal for many applications, or the sample size is large, so often, these conditions are met. Nevertheless, there are times when the sample sizes are not relevant. That is true when getting sample points is expensive, which is often the case in many scientific studies. In general, testing the hypothesis becomes very difficult. However, an important special case has been used extensively, namely when the underlying population is normal (but of unknown variance). We use Student's t-distribution. When we have a sample size n of a normal distribution with a mean μ and an unknown variance σ^2, we will review

$$t = \frac{\bar{x} - \mu}{s/\sqrt{n}} \tag{4.16}$$

Here, s is the standard deviation of the sample and compare these using $n - 1$ degrees of freedom with Student's t-distribution. If $n > 30$, we can equate it to the regular normal distribution based on the central limit theorem.

4.4.2.3 Test of Proportions

We also come across data that correlate to discrete distributions rather than the continuous ones we observe. One instance of this is binomial distribution, which is useful for instances where events that we are examining have only two possible outcomes (such as "heads" or "tails" with a coin). In this case, we look at the number of successes (often called x) in n trials (leaving $n - x$ failures) and either look at the number of successes or look at x/n, the ratio of positive trials. The two are also synonymous.

The use of the p-value to determine whether or not to reject the null hypothesis is also more useful for this sort of test. Stated simply, the p-value is the probability that the random variable, assuming the null is valid, will take on values that are far from the mean. It is equivalent to looking at values within the critical region; if the likelihood of such an observation occurring is less than α, we dismiss the null hypothesis. The use of the p-value to determine whether or not to reject the null hypothesis is also more useful for this sort of test. Simply stated, the p-value is the probability that the random variable, assuming the null is valid, will take on values that are far or far from the mean. It is analogous to looking at values within the critical region; if the probability of such an event occurring is less than α, we reject the null hypothesis. It is much easier to work with p-values and accept or reject the null hypothesis based on the condition $z_0 < \alpha$ where:

$$z_0 = \frac{|p_1 - p_2|}{\sqrt{\left(\frac{x_1 + x_2}{n_1 + n_2}\right)\left(1 - \frac{x_1 + x_2}{n_1 + n_2}\right)\left(\frac{1}{n_1} + \frac{1}{n_2}\right)}} \tag{4.17}$$

where n_1 = size of sample #1, n_2 = size of sample #2, x_1 = number of elements in sample #1 in category of interest, x_2 = number of elements in sample #2 in category of interest and:

$$p_1 = \frac{x_1}{n_1}, \ p_2 = \frac{x_2}{n_2}$$

4.4.2.4 Test of k-Proportions or Test of Independence

Let us expand to explore K proportions rather than just one. Such methods would help us to see if observed variations in proportions are due to chance or if the variations are significant. Suppose we have $X_1, \ldots, X_K$ independent random variables with π_i probability of success in n_i number of trials in a binomial distribution. Thus, we have populations of K, and exactly one proportion is correlated with each population. Let $x_1, \ldots, x_K$ be the values observed from those distributions. If all n_i's are large enough, then we can use the central limit theorem to approximate each one by the normal standard:

$$z_i = \frac{X_i - n_i \pi_i}{\sqrt{n_i \pi_i (1 - \pi_i)}} \tag{4.18}$$

We know that because each one gives a normal distribution:

$$\chi^2 = \sum_{i=1}^{K} \frac{(x_i - n_i \pi_i)^2}{n_i \pi_i (1 - \pi_i)} \tag{4.19}$$

It will be a χ^2 distribution with K degrees of freedom. Therefore, to test the null hypothesis that $\pi_1 = \pi_2 \cdots = \pi_K = \pi_0$ against the alternative hypothesis that some $\pi_i \neq \pi_0$, we use $\chi^2 \geq \chi^2_{\alpha, K}$ as our critical region of size α.

4.4.2.5 Summary of Hypothesis Testing

In general, the following protocol is used for hypothesis testing:

Step 1 State the Hypothesis to Be Tested

Two Tail	Upper Tail	Lower Tail
$H_0 : \mu_1 = \mu_2$	$H_0 : \mu_1 \leq \mu_2$	$H_0 : \mu_1 \geq \mu_2$
$H_1 : \mu_1 \neq \mu_2$	$H_1 : \mu_1 > \mu_2$	$H_1 : \mu_1 < \mu_2$
or	or	or
$H_0 : \pi_1 = \pi_2$	$H_0 : \pi_1 \leq \pi_2$	$H_0 : \pi_1 \geq \pi_2$
$H_1 : \pi_1 \neq \pi_2$	$H_1 : \pi_1 > \pi_2$	$H_1 : \pi_1 < \pi_2$

Step 2 Select an α value for significance 0.05, 0.01, or 0.001

Step 3 Obtain the information from the samples collected, i.e. $n_1, \bar{x}_1, s_1, n_2, \bar{x}_2, s_2$
Furthermore, compute the test statistics as appropriate.

$$t_0 = \frac{\bar{x}_1 - \bar{x}_2}{s_p \sqrt{\frac{1}{n_1} + \frac{1}{n_2}}}, \text{ where} \qquad z = \frac{\bar{x} - \mu}{\sigma / \sqrt{n}} \qquad z_0 = \frac{|p_1 - p_2|}{\sqrt{\frac{(x_1 + x_2)}{(n_1 + n_2)}\left(1 - \frac{x_1 + x_2}{n_1 + n_2}\right)\left(\frac{1}{n_1} + \frac{1}{n_2}\right)}},$$

$$s_p = \sqrt{\frac{(n_1 - 1)s_1^2 + (n_2 - 1)s_2^2}{n_1 + n_2 - 2}},$$

Step 4 Use t-distribution percentiles with $n_1 + n_2 - 2$ degrees of freedom to estimate the tail area beyond the upper limit of t_0. Or use the Z distribution if $(n_1 + n_2) = 30$ or higher

Step 5 For a two-tailed test, $p =$ two times the area in Step 4.
For a one-tailed test, $p =$ area in Step 4.

Step 6 If $p < \alpha$, conclude H_1 with $(1 - p)100\%$ confidence.
If $p \geq \alpha$, fail to reject H_0.

CHAPTER 5

Design of Experiments and Analyses

TODAY, EXPERIMENTS ARE CARRIED out in many science and engineering research fields to enhance our understanding and awareness of different scientific concepts and processes. Such studies are mostly carried out in a series of trials or evaluations, generating quantifiable results. The quantitative and qualitative effects of these experiments or trials are assessed using analytical instruments and measuring equipment. Scientific instruments/devices such as mass spectroscopy, atomic force microscopy, and electron microscopy are often used to measure the experiments' quantifiable results. The changes or variations found in the result of an experiment may occur from two sources: (1) The direct effect of the input variable's deliberate shift as the outcome variable is associated with the changing input variable and (2) the error or variance that stems from the measuring device and system.

Several factors affect an instrument's efficiency or measuring equipment, including the natural variability of the instrument/equipment and the human factor. When dealing with science, materials, and devices on a nanoscale and operating the measuring equipment close to or at the performance limit, it becomes critically important to quantify all variability sources to draw correct logical conclusions.

DOI: 10.1201/9781003206743-5

5.1 MEASUREMENT SYSTEM ANALYSIS

Many numerical claims are precise; however, there is a degree of uncertainty in all measurements that can stem from different sources. The process of assessing the uncertainty associated with a measurement is often called an analysis of uncertainty or measurement system analysis (MSA). The full description of a calculated value will estimate the confidence level associated with that value. Proper reporting of an experimental outcome and its uncertainty enables other people to decide the experiment's validity and encourages accurate comparisons with other related values or a theoretical forecast. This process is fundamental for a scientific hypothesis to be confirmed or rejected.

If we do a measurement, we usually believe that there is a particular exact or actual value depending on how we interpret what is being measured. Although we can never really know this actual value, we focus on finding this perfect quantity with the time and money available to the best of our ability. When we do measurements using different methods, or even when we do multiple measurements using the same method, we can get slightly different outcomes. Moreover, how do we document our findings on this mysterious real value to our best estimate? The most prevalent way is showing the range of values we think including the correct value. The MSAs identify and quantify the different variation sources that affect a measurement system. However, before we investigate the MSAs, we need to define the precision and accuracy.

5.1.1 Precision and Accuracy

Any discussion of error is commonly associated with two terms: "precision" and "accuracy." Precision refers to a measurement system's ability on an average to reproduce a measured value again and again. In other words, the precision measures the reproducibility of the measurement system. At the same time, accuracy is a measure of its closeness to the actual value. The accuracy thus measures a measurement system's ability to produce an average value that matches the actual accepted value. In other words, precision measures the repeatability of the measurement system. Sometimes together, they are also called measurement gauge's repeatability and reproducibility (or Gauge R&R). The series of targets in Figure 5.1 demonstrates the concepts of precision and accuracy.

If the "true value" is the target's core, A is precise and accurate, which is the laboratory's goal. The situation in B is precise (reproducible) but not accurate. The average marks of target C give an approximate result

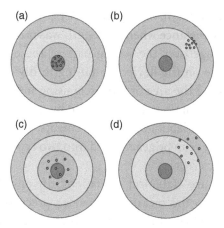

FIGURE 5.1 Visually (a) precise and accurate, (b) precise but not accurate (c) accurate but not precise, (d) neither accurate nor precise.

(accurate) but low precision. Target D demonstrates neither accuracy nor precision. Unfortunately, the terms error and uncertainty are often used interchangeably to characterize inaccuracy and imprecision. This use is so widespread that it is difficult to avoid it altogether. In order to evaluate the precision of a given measurement, the exact actual value must be known. Often, we have a calculated "textbook" value believed to be our "ideal" value and used to estimate the accuracy of our result. We can also use an abstract value that is measured based on universal values and theoretical principles and can be taken as an "ideal" value.

Nevertheless, physics is an analytical science, which implies scientific confirmation of the theory and not the other way around. We can solve these problems and retain a reliable accuracy sense by ensuring that we can compare our experimental value with the best-agreed value available even though we do not know the exact value. For example, if there is no agreed value to compare with, we might look up the manufacturer's precision requirements. However, all experiments do have some degree of error or uncertainty, no matter how meticulously planned and executed. In general, in labs, when gathering data or recording tests, one can learn how to recognize, correct, and analyze sources of error in an experiment and communicate the measurements' accuracy and precision.

5.1.2 Types of Errors

When scientists refer to experimental mistakes, they do not refer to what is generally referred to as accidents, blunders, or miscalculations. Sometimes referred to as illegal, human, or personal errors, specific errors may result from measuring the distance when the length was measured: alternatively measuring the voltage over the wrong part of the electrical circuit, or misreading the scale on the device, or failing to divide the diameter by two before calculating the area of the circle using formula $A = \pi r^2$. These errors are significant, but they can be avoided by experimenting again correctly the next time. In laboratory measurements, what scientists refer to as experimental errors can be generalized into two types of errors: random error and systematic error.

Random errors are statistical deviations (in any direction) in the measured data due to the measuring device's precision limitations and uncontrollable factors affecting the experimental effects. These include accidental errors caused by changing experimental conditions outside the experimenter's control; examples are equipment movements, humidity changes, and fluctuating temperatures. These errors may also include those that may be caused by humans and by accidental mistakes. Human errors include miscalculations in data processing, inaccurate reading of an instrument, or a personal bias in believing that some readings are more accurate than others.

Through their very nature, random errors cannot be quantified precisely because the magnitude of the random errors and their effect on the experimental values are different for each experimental repetition. So, statistical approaches are typically used to get an estimate of the experiment's random errors. Statistical analysis can measure random errors, which can be minimized by averaging over many measurements. Enough steps result in uniformly distributed data spread around an average or mean value. This positive and negative data dispersal is characteristic of random errors. The calculated standard deviation (the number of errors for a data set) is often recorded with measurements, as random errors are hard to remove. Also, to "smooth out" random error, a "best-fit line" is drawn through the graphed data.

Systematic errors: A systematic error is an error that occurs continuously in just one direction every time the experiment is conducted, i.e. the measurement value will always be larger (or lower) than the actual value. These errors are technical or procedural errors that trigger "lopsided" results, repeatedly deviating from the actual value in one direction. Examples of systematic errors: measuring a distance using the worn end of a meter stick, using an uncalibrated instrument, e.g. when a spectrometer drifts away from calibrated settings; a methodological error occurs when unintentionally neglecting the effects of viscosity, air resistance, and friction.

While in practice, it is difficult to predict the extent and severity of the systemic errors, some effort should be made to measure their effect wherever possible. Care should be taken in every experiment to remove as many structural and accidental mistakes as possible. Proper device configuration and adjustment can minimize the systemic errors leaving only the unintentional and human errors to cause any data spread. If, when calibrating against a standard, a systematic error is identified, applying a correction or correction factor to offset the effect can reduce the bias. Even though statistical methods will eliminate random errors, there is little use in reducing random errors below the measuring instrument's precision limit.

5.1.3 Error Estimation and Reporting

It is possible to quantify and report the exact contribution of all sources of variability in the measurement system. The following procedures, if followed correctly, can provide an excellent estimate of all sources of variability and their quantitative impact on the measurements. In order to perform a complete MSA, it is recommended to include the following:

1. Known standard samples cover at least 80% of the measurement range that the system must assess on a routine basis.

2. Suppose multiple sets of identical equipment are used, for example, if multiple rulers are used to measuring an object's length. Alternatively, multiple optical microscopes are used to assess the sample's size. At least two or more of these should be included in the MSAs to quantify the variability associated with using the equipment.

3. Similarly, since different people will be involved in experimentation, at least two to three experimenters should be involved in the MSAs.

4. The standard samples that will be used to assess the measurement systems variability should be blind-marked. The person measuring the sample should not know about the measurement's expected outcome to avoid the measurer's bias (human bias).

5. Each sample will then be measured multiple times (at least twice) by each experimenter using the same equipment. These measurements are then replicated for each set of equipment.

Let us assume that we have n standard samples to evaluate the measurement system's variability. We decide to use l experimenters that measure each sample m times and record the data. Let x_{ijk} denote the result of the measurement on the ith sample by *the jth* experimenter at the kth time, σ_{11} the standard deviation observed in sample 1 by experimenter 1 for all repetitions, and σ_{n1} the variability observed by experimenter 1 in sample n for all repetitions, up to σ_{nl}. Overall variability observed by experimenter 1 while measuring all samples across all repetitions, and variability observed in sample n by experimenter l for all repetitions, is given by $\sigma_{n1} = \sqrt{\sum_{i=1}^{n} \sigma_{i1}^2}$, and similarly $\sigma_{n2} = \sqrt{\sum_{i=1}^{n} \sigma_{i2}^2}$ and $\sigma_{nl} = \sqrt{\sum_{i=1}^{n} \sigma_{il}^2}$.

Furthermore, thus, the variability of the system affecting the system's ability to repeatedly produce the same value (variation obtained by the same person using the same equipment on the same sample for repeated measurements [variability within operator/device combination]) is given by:

$$\sigma_{\text{repeatability}} = \sqrt{\sum_{i=1}^{n}\sum_{j=1}^{l} \sigma_{ij}^2} \tag{5.1}$$

If $\bar{x}_1, \bar{x}_2, \ldots, \bar{x}_l$ are the averages of samples 1 to n for all repetitions by operators 1, 2, …, l respectively, then:

$$\bar{x} = \frac{1}{l}\sum_{j=1}^{l} \bar{x}_j \tag{5.2}$$

$$\sigma_{\text{reproducibility}} = \sqrt{\frac{\sum_{j=1}^{l}\left(\bar{x} - \bar{x}_j\right)^2}{l-1}} \tag{5.3}$$

Thus, the variability introduced by the measurement system can be approximated as follows:

$$\sigma_{\text{measurement}}^2 = \sigma_{\text{repeatability}}^2 + \sigma_{\text{reproducibility}}^2 \tag{5.4}$$

Total variability observed, $\sigma^2_{total} = \sigma^2_{samples} + \sigma^2_{measurement}$, is determined from mean $\overline{X}$ of all samples evaluated by all experimenters through all the replications as given by:

$$\overline{X} = \frac{1}{lmn} \sum_{i=1}^{n} \sum_{j=1}^{l} \sum_{k=1}^{m} x_{ijk} \qquad (5.5)$$

And the total variability is given by

$$\sigma^2_{total} = \frac{1}{(l-1)(m-1)(n-1)} \sum_{i=1}^{n} \sum_{j=1}^{l} \sum_{k=1}^{m} \left(\overline{X} - x_{ijk}\right)^2 \qquad (5.6)$$

A commonly used measurement system's capability measure is defined as precision to total ratio and is calculated as:

$$\frac{P}{Total} = \frac{\sigma_{measurement}}{\sigma_{total}} \qquad (5.7)$$

and this ratio must be below ≤0.10 or 10% for an adequate measurement system. Another very useful parameter to assess the capability is the ability of the measurement system to discriminate between two very closely spaced measurements and is known as the resolution Res. of the system and is given as:

$$Res. = \left(\frac{\sigma_{sample}}{\sigma_{measurement}}\right) \times 1.41 \qquad (5.8)$$

And this value should be higher than 5 for an adequate system. Suppose the system is inadequate based on the measurement in Equation 5.7 or 5.8 or both. In that case, the experimenters must investigate and improve their measurement system before launching the experimental study to derive any meaningful conclusion.

5.2 MULTIVARIATE REGRESSION ANALYSIS

Statistics are used for explaining the data and drawing inferences from the data in science. Inferential statistics are used to address questions about the data, to test hypotheses (formulate alternative or null hypotheses), to derive an impact estimate, usually a level or risk ratio, and to explain inter-actions (correlations) or model relationships (regression) within the data and in several other functions. Generally, estimated coefficients are the results of associations or the intensity of the effects.

Linear regression models offer a simplified approach to supervised learning. These models are easy but successful in modeling correlations. Recall that linear means that the variables are arranged in or extended along a straight or almost straight line. Linear implies that the relationship between the dependent and independent variables can be represented in a straight line. The line equation is $y = mx + c$. One dimension is the y-axis representing the dependent variable; the other dimension is the x-axis representing the independent variable. It can be plotted on a two-dimensional plane.

The generalization of this relationship may be expressed as $y = f(x)$. This generalization means defining the dependent variable as a function of the independent variable. If the dependent variable must be expressed in more than one independent variable, the generalized function is $y = f(x, z)$. In this case, we have got an additional dimension. We are trying to describe y as a combination of x and z. For a simple linear regression model, a straight line expresses y as a function of x. With an additional dimension (z), a plane will be required to express the linear relationship of dependent variable y with independent variables x and z.

Generalizing the linear regression equation, the two variable dependence can now be expressed as $y = m_1 x + m_2 z + c$; y is the dependent variable, i.e. the variable that needs to be estimated and predicted. x is the first independent variable, i.e. the variable that can be controlled. Variable x's slope, m1, defines the angle of the plane's xy-intercept line with the x-axis. Similarly, the second independent variable, z's slope, is m_2 that provides the yz-intercept line's angle to the z-axis, c is a constant that sets the value of y when both x and z are 0. A general linear model with two input variables may be expressed as:

$$y = \beta_0 + \beta_1.x_1 + \beta_2.x_2 \tag{5.9}$$

There can be many dimensions in the machine learning world. The generalized equation for the multivariate regression model in an n-dimensional space can be as follows:

$$y = \beta_0 + \beta_1.x_1 + \beta_2.x_2 + \ldots + \beta_n x_n \tag{5.10}$$

For performing analyses using inferential statistical tools, the natural variables (or the assigned arbitrary variables) are converted into dimensionless coded variables $x_{i1}, x_{i2} \dots x_{in}$ on a scale of -1 (the lowest value) to $+1$ (the highest value) as follows:

$$x_n = \frac{z_i - (z_{+1} + z_{-1})/2}{(z_{+1} - z_{-1})/2} \tag{5.11}$$

This equation converts all variables in a study to a common scale and evaluates each variable's impact and its interaction on the outcome independent of its measurement units. It is unlikely that one can make point estimates close to true ones while confounding; calculation errors, selection bias, and random errors are present. Random errors are not preventable in the estimation process. The simplest model of the response surface can be approximated by multiple linear regression as an extension of Equation 5.10 with "n" input variables as follows:

$$y_i = \beta_0 + \sum_{j=1}^{n} \beta_j x_{ij} + \varepsilon \tag{5.12}$$

where the linear regression coefficients, β_j, are estimated from the experimental data through the model fitting. When change in output variable due to a change in an input variable x_j depends on the values of other variables, this can be expressed as:

$$y_i = \beta_0 + \sum_{j=1}^{n} \beta_j x_j + \sum_{i=2}^{n} \sum_{j=1}^{i-1} \beta_{ij} x_i x_j + \varepsilon \tag{5.13}$$

Here the term $\beta_{ij} x_i x_j$ represents the statistical interaction between variable x_i and x_j. Both Equations 5.12 and 5.13 are linear terms and thus represent the first-order model. The relationship between variables may not always be linear, and a more complex model may need to be developed. If a strong quadratic relationship is expected, this can be expressed as a second-order model:

$$y_i = \beta_0 + \sum_{j=1}^{n} \beta_j x_j + \sum_{i=2}^{n} \sum_{j=1}^{i-1} \beta_{ij} x_i x_j + \sum_{j=1}^{n} \beta_{jj} x_j^2 + \varepsilon \tag{5.14}$$

Parameters b are often estimated from experimental data using computation software designed for this purpose and are donated by $\hat{\beta}$. The values predicted by Equation 5.12 for $\hat{y}_i$ is based on the input variables $x_{i1}, x_{i2} \ldots x_{in}$ are given as:

$$\hat{y}_i = \hat{\beta}_0 + \sum_{j=1}^{n} \hat{\beta}_j x_{ij} + \varepsilon \tag{5.15}$$

Similarly, the values predicted for Equations 5.13 and 5.14 based on input variables $x_{i1}, x_{i2} \ldots x_{in}$ are given as:

$$\hat{y}_i = \hat{\beta}_0 + \sum_{j=1}^{n} \hat{\beta}_j x_j + \sum_{i=2}^{n} \sum_{j=1}^{i-1} \hat{\beta}_{ij} x_i x_j + \varepsilon \tag{5.16}$$

And

$$\hat{y}_i = \hat{\beta}_0 + \sum_{j=1}^{n} \hat{\beta}_j x_j + \sum_{i=2}^{n} \sum_{j=1}^{i-1} \hat{\beta}_{ij} x_i x_j + \sum_{j=1}^{n} \hat{\beta}_{jj} x_j^2 + \varepsilon \tag{5.17}$$

The difference between observed and predicted values is called residual and is useful in making inferences about the model's adequacy. The residuals are calculated as:

$$\hat{\varepsilon}_i = y_i - \hat{y}_i \tag{5.18}$$

5.3 ANALYSIS OF VARIANCE

The simple linear model and the variance (ANOVA) model analysis can be a particular case of a more general linear model. The variations of one variable y are described by n explanatory variables x_i. The least square fitting is an actual application of this. The idea is to estimate y by a linear combination of x_i, such that the model is the best fit of y in the least squared sense. The approach used to check this problem is to measure (or obtain) the total variation and decompose it into variation sources.

The exploratory, experimental data analysis has a crucial drawback. The drawback is that the data are typically only analyzed in "slices." For example, we can look at the pattern due to changes in a specific variable (and residual variability) by ignoring blocks and then looking at the pattern.

However, it is not easy to look at multiple pattern components at the same time. This pattern recognition is significant when several variables or data structures are complex, typically the case for large historical data sets. We need a method of assigning variability to various sources at once: This is the function of ANOVA. The results taken from ANOVA are still arbitrary. Some measure of precision must be attached to the individual effects observed. For this reason, we need more systematic methods of study, and a hypothesis test (t-test in normal data) is typically used for individual results.

A total sum of square residues for a sample size of "N is used to measure variance in the method and is given as:

$$\text{Variance} = s^2 = \frac{\sum_{i=1}^{n} \varepsilon_i^2}{N-1} = \frac{\sum_{i=1}^{n} (y_i - \hat{y}_i)^2}{N-1} \tag{5.19}$$

Let us consider the example of a large set of data on some experiments with a sample size "N" that we have collected. Within this data set, there are two groups (two levels), one group of "K" samples where we used material "A" while the other group of (N-K) samples we used material "B," each with their own normal distributions. In this case, two more square residuals can be calculated for making inferences; that is, the sum of squares between the two materials' distribution.

$SS_B = \sum_{i=1}^{n} n_i (\hat{y}_i - \hat{y})^2$ and the sum of squares within the distribution from any given material $SS_W = \sum_{i=1}^{a} \sum_{j=1}^{n} (y_{ij} - \hat{y}_i)^2$ and $SS_E = \sum_{i=1}^{n} (y_i - \hat{y}_i)^2$.
The total sum of squares for the overall distribution containing all experimental data is thus given as:

$$SS_T = \sum_{i=1}^{a} \sum_{j=1}^{n} (y_{ij} - \hat{y})^2 = SS_B + SS_W + SS_E \tag{5.20}$$

where a = the number of groups/levels, y_{ij} = the jth data point in the ith group, n_i = the number of data points in the ith group or level, $\hat{y}$ = the grand mean, and $\hat{y}_i$ = the mean of the ith group or level. Mean square variance total (MS_T), the mean square variance between (MS_B), and mean square variance within (MS_W) can be calculated from Equation 5.20 by dividing the sum of squares with the appropriate degrees of freedom:

$$MS_T = \frac{\sum_{i=1}^{a}\sum_{j=1}^{n}\left(y_{ij}-\hat{y}\right)^2}{N-1}, MS_B = \frac{\sum_{i=1}^{a}\sum_{j=1}^{n}\left(y_{ij}-\hat{y}\right)^2}{K-1},$$

$$MS_W = \frac{\sum_{i=1}^{a}\sum_{j=1}^{n}\left(y_{ij}-\hat{y}\right)^2}{N-K}$$

$$(5.21)$$

If between-the-group variation is more extensive than within-the-group variation (i.e. $MS_B > MS_W$), the most variation in y is x. However, if between-the-group variation is smaller than the within-the-group variation (i.e. $MS_B < MS_W$), variation in y is due to things other than x. The ANOVA determines whether differences between averages of the levels are more significant than those that could reasonably be expected from variation within a level. This determination could be extended to include multiple variables divided into blocks, and ANOVA analyses could test the validity of the model and the confidence level therein. F-statistic (critical value for F) and the model correlation coefficient R^2 are thus calculated from Equations 5.20 and 5.21 as:

$$F_0 = \frac{MS_B}{MS_W} \ \& \ R^2 = \frac{SS_B + SS_W}{SS_T} = 1 - \frac{SS_E}{SS_T} \qquad (5.22)$$

F-statistics could be used to calculate the confidence interval by accepting or rejecting the assumption that the overall medal is valid. Thus, the correlation coefficient measures the percent total variability explained by the variables included in the model. When used to compare two averages or the average and known value (as in the case of individual variables used in the model), ANOVA simplifies the t-test based on the t-distribution.

Since the observed differences in response can be either due to the real effects of the change in input variables or the artifacts of random variation, it is essential to distinguish between these two situations. The statistical approach using an at-statistics-based hypothesis test is the only objective way to draw such conclusions from the data. The t-distribution is asymmetrical probability distribution centered at zero, close to the normal probability distribution. The difference is that the t-distribution has a variance that depends on the degree of freedom of the standard error in the statistics of interest.

The t-distribution is used to determine t-statistic that is then used to calculate confidence level for the true impact of the variable under consideration on the output or dependent variable and is given as:

$$t_0 = \frac{\bar{y}_1 - \bar{y}_2}{S_p\sqrt{\dfrac{1}{N_1} - \dfrac{1}{N_2}}}, \text{where pooled standard deviation}$$

$$S_p = \sqrt{\frac{(n_1 - 1)S_1^2 + (n_2 - 1)S_2^2}{n_1 + n_2 - 2}} \tag{5.23}$$

The region under the t-distribution beyond the critical value of t_0 is donated by α. The t-statistic assigns a confidence level $(1 - \alpha)$ to decision-making to include the variable in our model or exclude it from the model. In other words, we mean that there is a degree of trust that the variable in question has a statistically significant impact on the output variable or the property of interest. For example, at 90% confidence level $1 - \alpha = 0.90$ and $\alpha = 0.10$, conventionally used levels are 90% (somewhat confident), 95% (confident), and 99% (very confident).

Simply put, one tests the hypotheses $H_0: \beta_j = 0$ vs. $H_1: \beta j \neq 0$. Unless the null hypothesis, H_0, is not rejected, there is no proof that the values of the input variable x_j result in systematic differences in response variable y and that the term can be omitted from the model. On the other hand, if H_0 is rejected, the corresponding term has a statistically significant impact and should be retained in the model.

5.4 EXPERIMENTAL DESIGNS

Experiments have a central place in science, particularly nowadays, due to the complexity of science's problems. Generally, however, the scientific research is organized and carried out chaotically. There is, therefore, a question of the efficiency of the use of the experiment. The most common experimental technique applied by many modern scientists is one-variable-at-a-time (OVAT). One variable is changed at a time, leaving all the other variables unchanged (or constant) in the experiment, and the corresponding outcome shift is examined. Such experiments' effective efficiency is estimated, and the coefficient of their usability is around 2%.

To understand this point better, let us consider the example of an experiment. We know that the strength of metal matrix composites depends on the processing conditions. We have narrowed it down to two parameters in the processing conditions that we would like to experiment with: processing temperature and time. In the traditional way of experimenting, we would fix the processing time. We would vary the temperature to find the optimum processing temperature to improve the material's strength and produce a plot shown in Figure 5.2a.

Then we would run a second experiment where we would fix the temperature at the value we found in the first experiment for the temperature that maximizes the strength and then varies the processing time in this second experiment to produce Figure 5.2b.

What we have done in these two experiments is traversed the temperature–time design space along two lines (a vertical and a horizontal line), as shown in Figure 5.2c.

Suppose the objective of our experimentation is to maximize the strength of our material. In that case, we can only find the maximum from these experiments if, by luck, we choose the values of temperature and

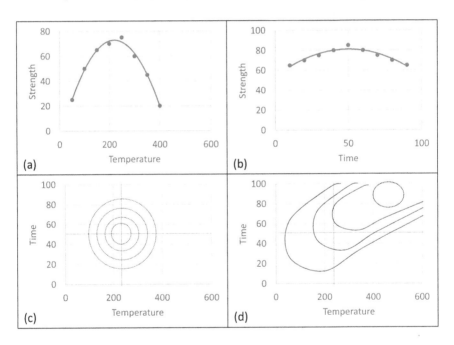

FIGURE 5.2 (a) Temperature vs material strength graph for the first experiment. (b) Time vs strength graph from the second experiment. (c) Demonstration of the time–temperature design space, if strength is concentric with our choice of values for experimentation. (d) If the optimum is not concentric with the time–temperature design space, lot more experimentation will be required to hit maximum strength.

time to experiment with that are concentric with the strength contours that are circular in the time–temperature design space. However, suppose the material strength follows noncircular behaviors and produces strength contours as shown in Figure 5.2d.

In that case, our effort at finding the maximum strength through this type of experimentation may end up costing us an infinite number of attempts. To increase research efficiency, something completely new needs to be introduced into traditional experimental research.

One form of innovation is the use of the statistical design of the experiments (DOE). Statistical design of experiments is a structured approach to assessing the relationship between cause and effect and helps us to achieve the following:

- Minimization of the total number of experiments.

- Simultaneous variance across all factors.

- A simple technique that allows effective solutions to be identified after each series of experiments.

The technique for designing experiments has contributed to an expansion in the resolution of very complex problems in all fields of human activity. We will explain in detail the different experimental designs, their properties, construction, and analysis. We will start with straightforward designs and then move on to more complicated designs. Each design is based on a specific rationale and applies to certain experimental situations. There are, however, some simple, universal concepts of experimentation and experimental design that need to be clearly understood. Such concepts apply to the formulation of the question under review, the choice of the experimental design, the experiment's execution, the analysis of the data, and the findings' interpretation.

It is not generally understood that the way data are collected significantly affects how easy or how difficult it would be to evaluate and interpret the data after it has been collected. "Construction of Experiments," or DOE, is the methodology that discusses the issue of how to gather data most effectively and efficiently. DOE is a purposeful variation of the independent variable (inputs) to detect the resulting changes in dependent variables (outputs).

The DOE illustrates how best to deliberately change independent variables (input factors) that impact the property of interest, the dependent variable (the output factors). By calculating and evaluating the output, we can evaluate how – or even if – the output is influenced by a specific input. In layman's terms, DOE provides the best way for the experimenter to investigate a process and discover the underlying knowledge and continue to follow the pursuit of more knowledge, as illustrated in Figure 5.3.

DOE allows one to test multiple input factors (variables) simultaneously in the most efficient way possible. Although the variables are changed simultaneously, each input variable's effects can be evaluated independently of all other variables using sophisticated statistical analytical techniques. As shown in Figure 5.4, we can evaluate what variables contribute to a shift in the mean output and how much.

We can also evaluate what variables can reduce the variation in output variable (property) of interest and by how much. Similarly, DOE helps us identify and quantify the variable that can help us move the mean output and reduce the variation in the output of interest. Finally, it will also help us identify what variables have no impact on the output. This finding

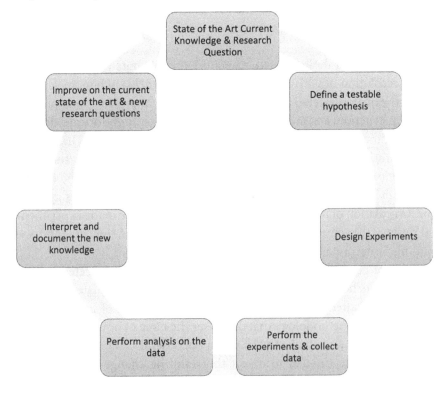

FIGURE 5.3 The process of knowledge development through the use of statistical design of experiments.

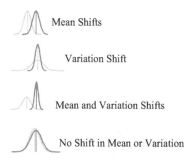

FIGURE 5.4   Knowledge gained through statistical design of experiments.

may also be advantageous, especially in an industrial setup where enormous efforts are made to control certain variables at the great expense of resources to produce an output desired by the market. Some of the experimenters refer to DOE as multivariate testing. Without DOE, engineers, researchers, scientists, or experimenters will not be as useful as this powerful but still unknown tool. While DOE is different from conventional tools and techniques, it is not difficult to learn or use.

The recent industrial popularity of DOE is related to the work of Taguchi, a Japanese engineer who concentrated on practical use over mathematical perfection of the technique. In short, Taguchi's research has started a revolution in the presentation of DOE material, where mathematical theory is downplayed to improve the clarity and practicality of the topic. As a result, scientists, engineers, technicians, and executives who are not mathematical experts are now becoming experimental design users. However, close inspection of the Taguchi method exposed shortcomings that contributed to the most recent evolution of the DOE, a blended approach of Taguchi and classical techniques.

To comprehend DOE, consider a field of physics as a phenomenon or process within nature. Comprehensive or complete knowledge of any aspect of this phenomenon is understood only by nature, although scientists typically have only a subset of that knowledge (Figure 5.5).

The difference between the present level of understanding and the ideal understanding of nature continues to expand as scientists today try to understand more complex processes. The typical approach to developments in physics has been to spend years of research trying to narrow the gap by using theoretical knowledge supplemented by OVAT experimentation. On the other hand, DOE will allow the gap to be quickly narrowed through proper preparation, design, data collection, review, and confirmation.

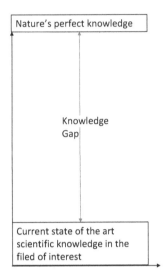

FIGURE 5.5   Visual representation of the scientific efforts and knowledge gap.

5.4.1 Two-Level Full Factorial Design and Analysis

Consider nature's absolute knowledge of any process where there is a real mathematical relationship that can explain process outputs as a function of all process inputs, e.g. Newton's second law of motion $F = ma$. Suppose nature knows this relationship, but we do not. How can we develop a reasonable understanding of this relationship quickly without getting tied up in years of theoretical research like Newton and his contemporaries did? A simple DOE for two input factors, such as mass (m) and acceleration (a), each evaluated at two levels, can be set up as four low (Lo) and high (Hi) experimental run (or trial) settings shown in Table 5.1.

The DOE set of experimental conditions is used to query the existence of underlying principles under such circumstances. The basic set of

TABLE 5.1   Two-Factor Two-Level Full Factorial Design

Trials	Input Factors and their Settings	
	Mass (m)	Acceleration (a)
1	Lo	Lo
2	Lo	Hi
3	Hi	Lo
4	Hi	Hi

conditions (or design matrix) is structured to evaluate the resulting data to construct a model to approximate the actual model found in nature (i.e. approximate the complete knowledge of nature). The layout in Table 5.1 is a full factorial experimental design of two variables that are tested at two levels. Many other experimental design styles are available to satisfy the goals, e.g. linear modeling, nonlinear modeling, screening. Assume that our experiment produced an average collection of results for the observed force "*F*" for each experimental run shown in Table 5.2.

The low and high levels of mass "*m*" (5, 10) and acceleration "*a*" (100, 200) are selected based on the experimenter's range of interest.

The DOE modeling analysis is typically conducted on homogeneous input values (that are dimensionless) to maximize the amount of information gained, with low input settings coded (−1) and high input settings coded (+1). These coded values will standardize the scale and the input variable units. The result is a new matrix of encoded inputs shown in Table 5.3. Note the addition of column *m*a* (mass*acceleration) in Table 5.3, which is created by the product of the columns *m* and *a* coded. This additional column is typically used to analyze the interactive or combined effect of *m* and *a*. Columns *m*, *a*, and *m*a* reflect the three variables whose effects are to be evaluated, i.e. the linear effects of *m* and *a* and the interaction effects of *m* with *a* (*m*a*). To get the data in the last three rows of Table 5.3, we have to do the following for each of the columns *m*, *a*, and *m*a*: find the average output when the respective effect column's value is (−1). Furthermore, similarly, find the average output when the respective effect column's value is (+1) and then find the difference between the two average effects (Δ). For example, when the *m* impact column is (−1) (trials 1 and 2), the output is 1,500 and 3,000, which averages out to 2,250.

TABLE 5.2   Two-Factor Two-Level Full Factorial Design with Actual Input and Output Values

	Input Factors and Their Settings		Output Factor
Trials	Mass (m)	Acceleration (a)	Average Force (F)
1	10	150	1,500
2	10	300	3,000
3	20	150	3,000
4	20	300	6,000

TABLE 5.3 Two-Factor Two-Level Full Factorial Design with Coded Input and Interaction Codes

Trials	Input Factors and Their Coded Settings		Interaction Term Coded	Output Factor
	Mass (m)	Acceleration (a)	m*a	Average Force (F)
1	−1	−1	+1	1,500
2	−1	+1	−1	3,000
3	+1	−1	−1	3,000
4	+1	+1	+1	6,000
Average (+1)	4,500	4,500	3,750	$\overline{F} = 3,375$
Average (−1)	2,250	2,250	3,000	
Δ = **Ave**(+1) − **Ave**(−1)	2,250	2,250	750	

The data in the last three rows of Table 5.3 are used to generate a DOE least squares predictive model for the force $\hat{F}$ as follows:

$$\hat{F} = \overline{F} + \frac{\Delta_m}{2}m_c + \frac{\Delta_a}{2}a_c + \frac{\Delta_{ma}}{2}m_c a_c \qquad (5.24)$$

where $\hat{F}$ is the predicted average force, $\overline{F}$ is the grand average of the experiments. Δ_m, Δ_a, and Δ_{ma} are the sizes of the linear effects of m, a, and combined effect of $m*a$, respectively, m_c is the coded variable for mass, and a_c is the coded variable for the acceleration. Substituting the values from the last three rows into Equation 5.24, we obtain our actual model as:

$$\hat{F} = 3375 + 1125m_c + 1125a_c + 375m_c a_c \qquad (5.25)$$

The prediction model presented in Equation 5.25 is for the coded values of the variables (i.e. +1 to −1 scale) for the variables m, variable a, and the interaction factor $m*a$. This model can be transformed into actual values using the following equation:

$$x_a = \left(\frac{x_H + x_L}{2}\right) + \left(\frac{x_H - x_L}{2}\right)x_c \qquad (5.26)$$

where x_a is the actual setting value for the variable of interest, x_c is the coded value of the variable of interest, x_H and x_L are the experimental high and low settings of the variable of interest. Using Equation 5.26, we can get the actual values for mass m_a and acceleration a_a as $m_c = \dfrac{m_a - 15}{5}$ and

$a_c = \dfrac{a_a - 225}{75}$, and substituting these for the coded values into Equation 5.25, we get:

$$\hat{F} = 3375 + 1125\left(\frac{m_a - 15}{5}\right) + 1125\left(\frac{a_a - 225}{75}\right) + 375\left(\frac{m_a - 15}{5}\right)\left(\frac{a_a - 225}{75}\right) = m_a a_a$$

$$(5.27)$$

In the above case, using a two-factor two-level full factorial design, we have produced the same model as Newton since our F data did not contain any errors or noise (friction, for example). There is noise or error to experimental data, and thus the DOE models can only be approximations to the actual relationship of nature. Nevertheless, imagine how easily Newton might have narrowed the gap if he had DOE in his scientific toolbox! Since DOE was not created until the 1900s, Newton never had access to the sophistication.

By a full factorial design, we mean that all possible combinations of the factors' levels are investigated in each complete test or replicate of the experiment. Considering the above example, suppose mass m and acceleration a are crossed; as we observed earlier, the effect of m or a is defined as the change in the force produced due to a change in the factor's level m or a. The entire design space is shown in Figure 5.6. This design is a 2×2 factorial design in which the levels of each factor interact with the levels of the other factor. However, depending on the study goal, the number of variables and their levels can vary, raising the number of experimental combinations and the logistics that follow.

Consider an experiment where we have a process that we are interested in investigating. The property of interest or the output variable, y, is virtually

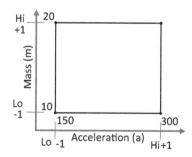

FIGURE 5.6   Two-factor two-level full factorial design space.

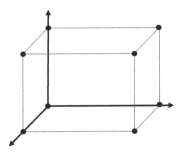

FIGURE 5.7   Three-factor two-level full factorial design space.

affected by three factors: x_1, chemical composition; x_2, temperature; and x_3, pressure. This design will be a three-dimensional design space with two levels and will have a total of eight runs, as shown in Figure 5.7. In other words, we shall have $2^3 = 8$ runs as listed in Table 5.4. The results of all eight sequences in the analyzed example serve to determine the factor effects, with seven tests being independent possibilities for testing the effects and one being used to compare them with the chosen fixed values. Three of the seven independently determined factor effects are used to assess the basic effect: Δ_{x_1}, Δ_{x_2}, and Δ_{x_3}, and the other four are used to assess their reciprocal interactions: $\Delta_{x_1x_2}$, $\Delta_{x_1x_3}$ $\Delta_{x_2x_3}$, and $\Delta_{x_1x_2x_3}$, following these expressions:

$$\Delta_{x_1} = \frac{y_5 + y_6 + y_7 + y_8}{4} - \frac{y_1 + y_2 + y_3 + y_4}{4} \tag{5.28}$$

$$\Delta_{x_2} = \frac{y_3 + y_4 + y_7 + y_8}{4} - \frac{y_1 + y_2 + y_5 + y_6}{4} \tag{5.29}$$

$$\Delta_{x_1} = \frac{y_2 + y_4 + y_6 + y_8}{4} - \frac{y_1 + y_3 + y_5 + y_7}{4} \tag{5.30}$$

$$\Delta_{x_1x_2} = \frac{y_1 + y_2 + y_7 + y_8}{4} - \frac{y_3 + y_4 + y_5 + y_6}{4} \tag{5.31}$$

$$\Delta_{x_1x_3} = \frac{y_1 + y_3 + y_6 + y_8}{4} - \frac{y_2 + y_4 + y_5 + y_7}{4} \tag{5.32}$$

$$\Delta_{x_2x_3} = \frac{y_1 + y_4 + y_5 + y_8}{4} - \frac{y_2 + y_3 + y_6 + y_7}{4} \tag{5.33}$$

$$\Delta_{x_1x_2x_3} = \frac{y_2 + y_3 + y_5 + y_8}{4} - \frac{y_1 + y_4 + y_6 + y_7}{4} \tag{5.34}$$

TABLE 5.4 Three-Factor Two-Level Full Factorial Design with Coded Input and Interaction Codes

Trials	Input Factors and Their Coded Settings			Interaction Term Coded				Output Factor
	x_1	x_2	x_3	x_1x_2	x_1x_3	x_2x_3	$x_1x_2x_3$	y
1	−1	−1	−1	+1	+1	+1	−1	y_1
2	−1	−1	+1	+1	−1	−1	+1	y_2
3	−1	+1	−1	−1	+1	−1	+1	y_3
4	−1	+1	+1	−1	−1	+1	−1	y_4
5	+1	−1	−1	−1	−1	+1	+1	y_5
6	+1	−1	+1	−1	+1	−1	−1	y_6
7	+1	+1	−1	+1	−1	−1	−1	y_7
8	+1	+1	+1	+1	+1	+1	+1	y_8

Like Equation 5.24, the values for the effects obtained from Equations 5.28 to 5.34 can feed into the development of a prediction model for the output of interest $\hat{y}$ as follows:

$$\hat{y} = \frac{\sum_{i=1}^{8} y_i}{8} + \frac{\Delta_{x_1}}{2}x_1 + \frac{\Delta_{x_2}}{2}x_2 + \frac{\Delta_{x_3}}{2}x_3 + \frac{\Delta_{x_1x_2}}{2}x_1x_2$$

$$+ \frac{\Delta_{x_1x_3}}{2}x_1x_3 + \frac{\Delta_{x_2x_3}}{2}x_2x_3 + \frac{\Delta_{x_1x_2x_3}}{2}x_1x_2x_3 \qquad (5.35)$$

which is like Equation 5.16 for $n = 8$.

5.4.2 Three-Level Full Factorial Design and Analysis

Although studies at two levels need fewer runs and are more economical to perform, practical considerations analyze factors with more than two levels desirable. For example, suppose we expected the relationships to be nonlinear and wanted to capture the relationships' curvature in the form of polynomials of various degrees. In that case, we will have to test the variables at at-least three levels.

A logical extension to three-level designs is the conceptual basis for two-level designs and their ramification for data analysis. The main difference comes from the fact that each three-level element has two degrees of freedom. There are two methods for parameterizing the interaction effects in three-level designs: the orthogonal component system and the linear quadratic system. Traditional variance analysis refers to the former method, while a new technique for regression analysis needs to be developed for the three-level system. To facilitate the fitting of a regression model, consider the simplest case: two variables, say A and B, each of the variables has three levels, denoted by 0, 1, 2. A variation of this $3^2 = 9$ factorial treatments is then defined by $x_i = (x_1, x_2)$, where $x_i = 0, 1, 2$ ($i = 1, 2$), with x_1 referring to factor A, and x_2 referring to factor B. A regression model relating the response "y" to x_1 and x_2 supported by a two-factor, three-level design is given by Equation 5.17. Notice that the addition of third level to factors allows the relationship between responses and the design factors to be modeled as quadratic.

A two-factor system tested at three levels would require $3^2 = 9$ experiments, as shown in Figure 5.8. This set of nine combinations of treatment can be divided into three sets of three combinations of treatment, each according to factor A levels as follows:

$$\text{Set I}: \{(0,0), (0,1), (0,2)\}$$

$$\text{Set II}: \{(1,0), (1,1), (1,2)\} \tag{5.36}$$

$$\text{Set III}: \{(2,0), (2,1), (2,2)\}$$

We can describe these three sets more formally utilizing the following three equations:

$$\text{Set I}: x_1 = 0$$

$$\text{Set II}: x_1 = 1 \tag{5.37}$$

$$\text{Set III}: x_1 = 2$$

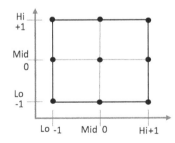

FIGURE 5.8   Two-factor three-level full factorial design space.

Comparisons with the mean proper outcomes are then assumed to contribute to the vital effect A for these three sets. Because there are three sets, these three sets contain two linearly independent comparisons (i.e. their mean responses), and these comparisons represent two degrees of freedom for primary effect A. The correlations may be (Set I−Set II) and (Set I−Set III), or (Set I−Set II) and (Set I+Set II−2 Set III). Likewise, we can divide the nine variations of treatment into three sets corresponding to the levels of factor B, equivalently, the equations:

$$\text{Set I}: x_2 = 0: \{(0,0), (1,0), (2,0)\}$$

$$\text{Set II}: x_2 = 1: \{(0,1), (1,1), (2,1)\} \tag{5.38}$$

$$\text{Set III}: x_2 = 2: \{(0,2), (1,2), (2,2)\}$$

Comparisons between the mean answers of these three sets then evaluate the main effect B. As in the two-level case, the interaction of factors A and B will be described in terms of set (of treatment combinations) comparisons, which are calculated by equations involving both x_1 and x_2. One such partitioning is given through:

$$\text{Set I}: x_1 + x_2 = 0: \text{mod } 3: \{(0,0), (1,2), (2,1)\}$$

$$\text{Set II}: x_1 + x_2 = 1: \text{mod } 3: \{(2,0), (0,2), (2,2)\} \tag{5.39}$$

$$\text{Set III}: x_1 + x_2 = 2: \text{mod } 3: \{(1,0), (0,1), (2,2)\}$$

Comparisons between these three sets represent two of the four degrees of freedom for the interaction between A and B. The remaining two degrees of freedom are accounted for by comparisons between sets centered on the partition below:

$$\text{Set I}: x_1 + 2x_2 = 0 \text{ mod } 3: \{(0,0), (1,1), (2,2)\}$$

$$\text{Set II}: x_1 + 2x_2 = 1: \text{mod } 3: \{(1,0), (0,2), (2,1)\} \tag{5.40}$$

$$\text{Set III}: x_1 + 2x_2 = 2: \text{mod } 3: \{(2,0), (0,1), (1,2)\}$$

To see what our design means in relation to the standard factorial representation so far, we consider:

$$\hat{y}_{ij} = \mu + A_i + B_j + (AB)_{ij} \tag{5.41}$$

Here, μ = overall mean of all the combinations of experiments. By $i, j = 0, 1, 2$ describing the factor levels A and B, two degrees of freedom account for the principal effects for A and B. Factors A and B interact with each other, accounting for four degrees of freedom, totaling eight degrees of freedom with nine experimental runs. The degrees of freedom can be further partitioned according to nature (i.e. qualitative or quantitative) of the variables. For simplicity, let us consider the case of both variables being quantitative, and let X_{il} and X_{2l} ($l = 0,1,2$) denote the levels of factors A and B evenly spaced, respectively. Then:

$$x_{il} = \frac{X_{il} - \overline{X}_{i.}}{\frac{1}{2}(X_{i2} - X_{i0})} \tag{5.42}$$

are the coded levels, with $x_{i0} = -1$, $x_{i1} = 0$, and $x_{i2} = +1$, ($i = 1, 2$ for the two factors). We can then write a model of the form:

$$y(x_{1l}x_{2l'})_m = \beta_0 + \beta_1 x_{1l} + \beta_2 x_{2l'} + \beta_{11} x_{1l}^2 + \beta_{12} x_{1l} x_{2l'}$$

$$+ \beta_{122} x_{1l} x_{2l'}^2 + \beta_{112} x_{1l}^2 x_{2l'} + \beta_{1122} x_{1l}^2 x_{2l'}^2 + \varepsilon(x_{1l} x_{2l'})_m \tag{5.43}$$

where $l, l' = 0,1,2$ and $m = 1,2,\dots r$. This equation is an explicit model that accounts for all degrees of freedom for two factors' main effects and interactions. Estimates of the regression coefficients can be obtained using the process of least squares. Hypothesis tests about these coefficients of regression can be performed. While this is clear, it is not always easy to understand because the estimators are correlated. Often, a representation in terms of orthogonal polynomials is more convenient. Let $P_0(x)$, $P_1(x)$, and $P_2(x)$ be the zeroth-, first-, and second-degree polynomials, respectively, then Equation 5.43 can be rewritten as:

$$y(x_{1l}, x_{2l'}) = \sum_{ii'=0}^{2} \alpha_{ii'} P_i(x_{1l}) P_{i'}(x_{2l'}) + \varepsilon(x_{1l}, x_{2l'})_m \tag{5.44}$$

or this can be simplified to matrix notation as

$$y = X\alpha + \varepsilon \tag{5.45}$$

where y is the observation column vector, X is the design model matrix of known constants, the column vector of the regression coefficients is α, and ε is the error column vector. Since the columns of X are orthogonal to each other, X'X is a diagonal matrix, and it facilitates the estimation of regression coefficients as:

$$\hat{\alpha} = (X'X)^{-1} X'y \tag{5.46}$$

However, the estimation of the regression coefficients becomes very complex compared with the two-level experiments. As the number of factors increases, for example, just three factors at three levels would require $27 = 3^3$ runs, as shown in Figure 5.9. The complexity of the calculations also increases and requires sophisticated computing to help with the least square estimation of the regression coefficients. Several software tools have been developed in the market to help with these estimations. Minitab, DoE Pro, Design Expert, and JMP are a few examples of software choices. Generalizing from these discussions on experiments, suppose we have n variables that we would like to test at k levels. Then our design space will require k^n trials (not including the repetition we would like to perform to improve our experimental outcomes' confidence level).

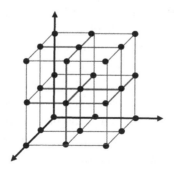

FIGURE 5.9   Three-factor three-level full factorial design space with 27 runs.

5.4.3 Partial Factorial or Fractional Factorial Designs

As the number of variables in a 2k factorial design increases, most experimenters' resources quickly outgrow the number of runs needed for a full replication of the design. A full replicate of the 26 design includes 64 runs, for example. Only 6 of the 63 degrees of freedom are used in this design to estimate the essential effects, and only 15 degrees of freedom are used to estimate the interactions of two variables. With three-factor and higher interactions, the remaining 42 degrees of freedom are correlated. Fractional factorial experiments are essential alternatives to complete factorial experiments if the implementation of complete factorial experiments is prevented by budgetary, time, or experimental constraints. Besides, studies involving several variables are routinely performed as fractional factorials. All possible factor combinations need not be evaluated to estimate the significant factor effects; usually, the essential effects and low-order interactions need to be estimated.

Suppose the experimenter can reasonably conclude those specific high-order interactions are negligible. In that case, knowledge about the main effects and low-order interactions can be obtained by running just a fraction of the entire factorial design. These fractional factorial models are among the industry's most used forms of design. Screening tests are an effective use of fractional factorials. These are studies in which several factors are considered to determine certain factors (if any) that have significant effects. Note that screening experiments are typically carried out early in a study when many of the variables considered initially are likely to have little or no effect. In subsequent studies, the variables that are established as significant are then explored in greater depth.

5.4.3.1 Half Fraction of 2k Design

Consider a situation in which three variables are of concern, each at two stages, but the experimenters cannot afford to run all $2^3 = 8$ combinations of treatment. However, they can manage four runs. This constraint indicates a half fraction of a configuration of 2^3. Since the design comprises $2^{3-1} = 4$ treatment variations, a 2^{3-1} design is also called a half fraction of the 2^3 design. In Table 5.4, the table of plus and minus signs for design 2^3 is shown. Suppose we choose the four combinations of treatments a, b, c, and abc as our one-half fraction. Table 5.4 can be rearranged to help understand the fractions a bit better, and the two half fractions of the design are reflected in Figure 5.10.

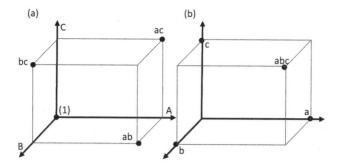

FIGURE 5.10 Two half-fractions of a three-factor two-level full factorial design space.

Note that by selecting only those treatment combinations with a plus in the ABC column, the 2^{3-1} template is created. ABC is thus called the generator of this unique fraction. We will generally refer to a generator like ABC as a term. The identity column I is always plus, so we call I = abc our defining relation for the fractional design. In general, the defining relationship for a factorial fraction will always be the set of all columns equal to the column I of the identity. The combinations of treatments in the 2^{3-1} system produce three degrees of freedom that we can use to predict the significant effects. Referring to Table 5.5, we note that the experiments used to estimate the critical effects of A, B, and C are linear combinations as follows:

$$[A]=\frac{1}{2}(a-b-c+abc)$$

$$[B]=\frac{1}{2}(-a+b-c+abc)$$

$$[C]=\frac{1}{2}(-a-b+c+abc)$$

where [A], [B], and [C] are used to denote the linear combinations of the key results. It is also easy to confirm that the linear combinations of the observations used to estimate the interactions of two variables are:

$$[BC]=\frac{1}{2}(a-b-c+abc)$$

$$[AC]=\frac{1}{2}(-a+b-c+abc)$$

$$[AB] = \frac{1}{2}(-a - b + c + abc)$$

Therefore, [A] = [BC], [B] = [AC], and [C] = [AB], so it is difficult to distinguish between A and BC, B and AC, and C and AB. We truly estimate A+BC, B+AC, and C+AB when we estimate A, B, and C. Two or more effects that have this property are known as aliases.

Now assume that we choose the other half fraction, that is, the treatment combinations associated with a minus in the ABC column in Table 5.5. Figure 5.10b indicates this alternative, or complementary, one-half fraction (consisting of the runs (1), ab, ac, and bc). For this design, the defining relation is $I = -ABC$. The linear combination of the observations from the alternate fraction, say [A]′, [B]′, and [C]′, gives us:

$$[A]' \rightarrow A - BC, [B]' \rightarrow B - AC \text{ and } [C]' \rightarrow C - AB$$

Thus, with this unique fraction, as we measure A, B, and C, we are merely estimating A−BC, B−AC, and C−AB. In practice, what fraction is used does not matter. The two fractions belong to the same family, i.e. the two half fractions form a complete full factorial design.

5.4.3.2 Orthogonal Array Designs or Taguchi's L Designs
Genichi Taguchi's work has had a significant effect on expanding interest in and use of planned experiments. For what he called robust parameter design, Taguchi encouraged the use of planned experiments to (1) making

TABLE 5.5 Rearranged Three-Factor Two-Level Full Factorial Design with Coded Input and Interaction Codes for Developing a Fractional Factorial

Trials	Input Factors and Their Coded Settings				Interaction Term Coded			
	I	a	b	c	ab	ac	bc	abc
A	+1	+1	−1	−1	−1	−1	+1	+1
B	+1	−1	+1	−1	−1	+1	−1	+1
C	+1	−1	−1	+1	+1	−1	−1	+1
AB	+1	+1	+1	+1	+1	+1	+1	+1
AC	+1	+1	+1	−1	+1	−1	−1	−1
BC	+1	+1	−1	+1	−1	+1	−1	−1
ABC	+1	−1	+1	+1	−1	−1	+1	−1
(1)	+1	−1	−1	−1	+1	+1	+1	−1

processes insensitive to environmental conditions that are difficult to control or other conditions, (2) making goods indifferent to component-transmitted variance, and (3) finding process variables at levels that push the mean to a target value while simultaneously reducing uncertainty around this value.

Taguchi proposed highly fractionated factorial designs and other orthogonal arrays, and some novel statistical methods to solve these problems. For these projects, Taguchi produced tables containing the test runs. The need to create the design and define the confounding pattern is removed by providing such tables. However, one must always be conscious of the design's resolution and, in some instances, the confounding patterns. Balance is achieved in an orthogonal array table since each level of a factor occurs with each level of each of the other factors an equal number of times. Note that all full factorials in which an equal number of repeats are orthogonal arrays for each factor-level combination. Some fractional factorial designs are orthogonal arrays, and others are not.

In the design of experiments, orthogonal arrays are highly attractive because they possess many substantial statistical properties, such as the statistical independence of estimated factor effects. There were orthogonal arrays of all the complete and fractional factorials provided in this chapter. Due to the importance put on them by Taguchi, who was instrumental in promoting their use using simple tables, some sets of orthogonal arrays have gained widespread acceptance and use in experimentation. The L8 array shown in Table 5.6 is an example of the ease with which tables of orthogonal arrays can be used in practical applications.

These designs are highly efficient in terms of the number of test runs. They are efficient if possible to perform fractional resolution III factorials

TABLE 5.6   Taguchi's L_8, Two-Level Fractional Factorial Orthogonal Design with Coded Inputs

	1	2	3	4	5	6	7
1	−1	−1	−1	+1	+1	+1	−1
2	−1	−1	+1	+1	−1	−1	+1
3	−1	+1	−1	−1	+1	−1	+1
4	−1	+1	+1	−1	−1	+1	−1
5	+1	−1	−1	−1	−1	+1	+1
6	+1	−1	+1	−1	+1	−1	−1
7	+1	+1	−1	+1	−1	−1	−1
8	+1	+1	+1	+1	+1	+1	+1

(immediate effects are unaliased only with other main effects). Numerous of the technically advocated orthogonal arrays consist of very few test runs. These design formats are also saturated so that the full number of variables possible can be used with up to one fewer variable than test runs. The primary design property that is often ignored is that the designs that result are of resolution III only. The existence of any influences on interaction biases the main effects. Orthogonal arrays for fractional factorial experiments of three levels with higher resolution are possible.

5.4.4 Historical Data Analysis

We live through times where data are all around us, and it is created, processed, analyzed, presented, and published, which again forms another kind of data for further study. Practical knowledge is released and reviewed in the scientific community. It is expected that the study will include a detailed overview of the results of the research contributions examined, analyze the current state of science, facilitate debates on research methodologies, and set the stage for ongoing and future research. The task is to turn independently collected technical knowledge produced into an alternative perspective utilizing individually conducted controlled perturbation experiments. The information produced may contain relations between a multitude of system variables undefined or not well understood.

Inferential statistical analysis of examined historical or experimental data may play a pivotal role in translating the observed data into essential knowledge to advance the underlying science and accelerating the achievement of the research goals in the field of interest. As outlined above, the most useful statistics are used in the statistical design of experiments. However, the problem is that the experimenter selects a set of experiments in a planned experiment and runs them as randomly as possible. There are two things this offers:

1. The matrix of the architecture ensures the variables of interest will be independent of one another.

2. Randomization of the runs also means that any unknown, uncontrolled variables' effects can only manifest themselves as contributing to the model's error word.

Suppose we take happenstance/historical data and try to develop a list of experiments that approximate what we do not know and cannot know about the experiments in the design space. This estimate includes the confounding effect of various unknown/uncontrolled variables associated with historical or experimental literature data runs. This estimate, in turn, means we will not know if the effect associated with a variable is associated with that variable and not some other variable or other combinations of variables – all of which are unknown and uncontrolled.

What we can do with historical data is the following: Take the block of data of interest, clean it by removing any apparent outliers, scale all of the X variables to a – 1 to 1 range, and evaluate the matrix of the X variables using the approach of eigenvalues and condition indices. Some of the software tools that have been built for statistical design of experiments, for example, DOE Pro and Design-Expert® software tools, have this option and will do the scaling and analyses for the experimenter when fed with clean data.

The program will then run a regression of all the X variables of interest against the output variables. With this method, for a given block of data, we can have a model expressing Y as a function of a group of X's. This model makes it possible for experimenters to use historical data that allow them possibilities of meaningful associations with the response of interest, Y. The approach could be used as the first step replacing the screening design to select the variables with the highest impact on the outputs of interest for a subsequent modeling design, provided adequate historical data are either available or collected from the published research papers.

5.5 SYSTEMS MODELING

Each treatment factor is defined in the actual experiment by various "levels": different types or different quantities, such as different processing times and different processing temperatures. Concerning other factors in the subject matter model, the investigator may decide to limit the factor to only one level, e.g. a single material composition only, or to include multiple levels, e.g. different material compositions. In the above case, these variables would also have to be considered in the subsequent data review. It is necessary to provide an adequate model of the response data for this reason.

There are various ways of interpreting data from such an experiment, just as there are different approaches to settling on a statistical design for an experiment. Taguchi's proposal to use different types of signal-to-noise ratios (S/N ratios) to evaluate and interpret such data has come under a certain amount of criticism. Other approaches have been suggested as a consequence of looking at the data from these experiments.

5.5.1 Linear Models

Linear regression is usually used to model one quantitative variable as a function of one or more other variables. Regression modeling is one of the most commonly used statistical modeling techniques to suit one or more predictor variables x_1, x_2, ..., x_p as a function of a quantitative response variable y. Regression models may be used to fit the data collected from a statistically constructed experiment in which either quantitative variables or factor level indicators are predictor variables. All the previously mentioned ANOVA models are particular forms of regression models. The predictor variables are specially coded indicator variables in ANOVA models, where the upper level of a factor is represented by $a+1$ and the lower level by $a-1$. Covariance analysis (ANCOVA) models are also unique forms of regression models in which predictor variables are some of the variables, and quantitative ones are others.

The use of nonexperimental data, such as empirical or historical data, is one reason for the widespread popularity of regression models. Another explanation is that regression analysis procedures provide diagnostic techniques for (1) detecting incorrect model parameters, (2) evaluating the effect of outliers on the fit, and (3) evaluating whether the fit is adversely affected by redundancies (collinearities) among the predictor variables. Maybe most pragmatically, regression models are commonly used because when the genuine functional relationship, if any, between the answer and the predictors is uncertain, they also offer excellent adaptations to a response variable. In addition to the random error variable, linear regression models are linear in the unknown parameters.

The coefficients in the model appear on the predictor variable (β_1) either as additive constants (β_0) or as multipliers. This criterion applies to various models of linear regression. Notice that, for example, $x = \ln z$ or $x = \sin z$ for any variable z, the predictor variable may be a component, linear or nonlinear, of other predictor variables. However, the predictor variable cannot be a function of unknown parameters, such as $x = \ln(z+\varphi)$ for any unknown φ, or $x = \sin(z-\varphi)$. Models that are nonlinear functions of parameters of an unknown model are referred to as nonlinear models.

The fitting of an answer to more than one predictor variable is involved in a multiple linear regression analysis. For example, as a function of several fuel properties, an experimenter might be interested in modeling vehicle emissions (y), including viscosity (x_1), cetane number (x_2), and distillation temperature (x_3).

Due to the possible interrelationships among the predictors, additional complexities are added in such analysis beyond the single-variable analyses. Two or more predictors may have synergistic effects on the response, as with ANOVA models, leading to the need to include the multiple predictors' joint functions to model the response accurately. Flexibility for the analyst is an advantage of modeling a response as a function of many predictor variables. A broader range of response variables can be satisfactorily modeled with multiple regression models than single-variable models. One is limited to using functions with only one predictor in a single-variable analysis. Different individual and combined functional types are enabled for each of the predictors in multiple regression analyses. Direct comparisons of alternative predictor decisions can also be made.

Studies of nonexperimental data such as these are also carried out to determine the reasons for the patterns. Extreme caution must be taken when evaluating experimental or nonexperimental data. Some of the variables are not monitored to ensure those accurate inferences are made when statistically relevant results are obtained. Confused predictor variables and spurious relationships must be concerned. Where the design factors or predictor variables simultaneously influence the values of two or more response variables, multivariate ANOVA models and multivariate regression models are appropriate. There are several different stages involved in a thorough regression analysis, each of which is important for the effective fitting of regression models. Those stages are:

- Plan: Plan the initiative for data collection.

- Investigate: Analyze the database; measure statistics for the summary; map the variables.

- Specify: Specify a functional type for each variable; formulate an initial model; and, if necessary, reexpress variables.

- Estimate: Approximate the model parameters; measure statistics that summarize the fit's adequacy.

- Assess: Examine the assumptions of the model; analyze diagnostics for impactful results and collinearities.

- Select: Choose predictor variables with statistical significance.

5.5.2 Nonlinear Models

Models displaying nonlinear relationships between the response and the predictor variables may be restated by acceptable response expressions or (a subset of) the predictor variables or both as a linear model. One benefit of such expressions is that they allow nonlinear relationships between a response and a collection of predictors to be added to the procedures developed for linear models. When the model is presumed to be linear, reexpression of the model may also be required. However, the errors do not have constant standard deviations (for example, when the magnitude of the standard error deviation is equal to the mean of the response), or the distribution of errors is considered to be nonnormal.

Nonlinear model respecification is not, however, always a viable choice. For the deterministic part of the model, reexpression of a nonlinear model might result in a complex linear form and the loss of the original measurement metric. Thus, the candidate model of choice is often nonlinear since it gives a more accurate explanation of the relationship between response and predictor variables. Likewise, nonlinear models are often the models chosen for practical purposes because of the problem's physical nature. For instance, the response of interest may be described in engineering experiments by applying a differential equation that is nonlinear in the model's parameters.

In some instances, to achieve an acceptable fit, one may also require the use of quadratic or cubic terms (say x^2, x^3). In other situations, with the use of transformation on the response itself, the curvature can be treated very easily. The experimental design used must contain adequate, separate design points. Besides, of course, to approximate the pure quadratic terms, the design must require at least three levels of each design variable.

The simplest model justified by what is known about the physical process being studied and by suggestions obtained from plots of the response variable versus the predictor variables should be started. Suppose a lack-of-fit test or a residual analysis shows that a proposed model is a poor approximation to the responses observed. In that case, the next higher-order terms can either be applied to the model or nonlinear models can be examined. A first- or second-order polynomial is adequate to characterize a response in many experimental situations.

CHAPTER 6

Basics of Machine Learning

A VAST AND FASCINATING EMERGING scientific field is machine learning that tempts many scientists to dive in and contribute. It has applications in various fields, such as gaming, from medicine to ad campaigns, from military to pedestrian tracking, from predicting congestion on highways to air traffic control, and much more. As more and more places look to it as a means of coping with the vast volumes of data available, its value continues to increase. Like knowledge, learning involves such a wide variety of mechanisms that it is impossible to define precisely. A description of a dictionary includes phrases such as "to gain intelligence, or comprehension, or skill of, by research, training, or practice" and "changing a behavioral propensity through practice." Concerning machines, we might conclude, quite generally, that a machine learns in such a manner when it changes its configuration, software, or data (based on its inputs or in response to external information). These improvements, such as adding a record to a database, fall easily into other disciplines and are not generally best known as studying. Nevertheless, for example, when a speech recognition system's output increases after hearing many recordings of a person's speech, we feel very confident in claiming that the system has improved in that situation.

With the aid of computers, the traditional solution to addressing issues is to write programs that address the problem. In this approach, the programmer must understand the problem, find a machine-friendly solution, and enforce this machine's solution. We call this method deductive since,

DOI: 10.1201/9781003206743-6

from the problem formulation, the person deduces the solution. However, in physics, genetics, chemistry, biophysics, medicine, and other research fields, a large amount of knowledge is generated that is difficult for humans to comprehend and interpret. The computer that learns can also find a solution to a problem. Such a computer analyzes the data and seeks structures in the knowledge automatically, i.e. learns. It is possible to use information about the extracted structure to solve the issue at hand. Machine learning (ML) is about inductively solving problems with machines, i.e. computers.

Researchers in scientific learning create algorithms that automatically enhance the solution to a problem with more data. In general, with the amount of problem-relevant data available, the efficiency of the solution improves. Problems solved by ML techniques range from the classification of data, the prediction of parameters, the organizing of data (e.g. clustering), the compression of data, the data visualization, the manipulation of data, the sorting of functional elements from data, the extraction of correlations between data components, the modeling of data generation systems, the creation of noise models for the observed data, and the aggregation of data from different sensors.

6.1 INFORMATION THEORY

Information theory is a field of applied mathematics that began with Shannon's paper "A Mathematical Theory of Communication," published in the Bell Systems Technical Journal in 1948. The paper explained how data could be quantified with absolute accuracy and highlighted all communication media's vital unity. Telephone messages, text, radio waves, and pictures could be encoded in bits. Any means of communication can be transformed into bits and communicated. Information theory significantly impacts ML. It is complicated and lacks more explicit definitions. There are some key concepts from the communication theory that need to be understood to comprehend its meaning entirely.

6.1.1 Channel Speed Limit

The most important of communication theory outcomes is the assumption that every communication medium has a speed limit, calculated in binary digits per second. The bad news is that error-free contact beyond the limit is mathematically unlikely. One cannot make the channel go faster than the limit without missing any information, no matter how sophisticated an error correction scheme one uses, and no matter how much one can compress the data. The good news is that zero-error data can be transmitted below this limit.

Of necessity, as the number of bits increases, it would be necessary to encrypt the files so that the majority of them get through and those that are missing can be recovered from the rest. Communication would become increasingly inefficient as the message's complexity and duration grew. However, one could make the likelihood of error as low as one wanted, below the cap.

6.1.2 Communication System Architecture

One of the significant achievements of "A Mathematical Theory of Communication" is illustrated in Figure 6.1: the architecture and design of communication systems. It shows that it is possible to separate any communication system into components, which can be treated as distinct mathematical models independently. Thus, the design of the source can be isolated from the design of the channel. This model has had applications in communication and computer machine theory, telephone exchange design, and other fields. All communication systems today are based on this model – it is a digital era blueprint.

In simple terms, the model indicates that the information transmission process starts with the source of information, through a transmitter to receiver and finally to the intended destination. In between, there are sources of noise that complicate the communication. A telephone line along which digital information is shared is an example of noisy communication networks – it suffers from cross-talking with other lines; the line circuitry distorts and adds noise to the signal transmitted.

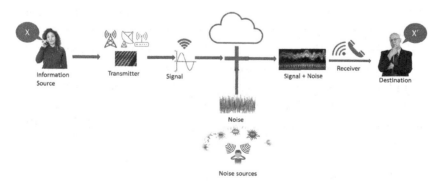

FIGURE 6.1 Communication system's architecture derived from the "A Mathematical Theory of Communication".

Galileo's radio communication relay, the space-orbiting Jupiter planet craft, receives background radiation from earthly and cosmic sources from a deep space network that listens to Galileo's trivial transmitter. The DNA is subject to mutations and damage in reproducing cells, in which the DNA of the daughter cells incorporates information from the parent cells.

The disk drive. This last example shows that contact does not have to require information traveling from one place to another. We will read it off in the same place, except when we write a file on a disc drive later. A disc drive writes a binary digit (a one or zero, also known as a bit) in one of the two directions by aligning a magnetic material patch. This disc may later fail to read the stored binary digit for any of the following three reasons: (1) The magnetization may spontaneously flip in a material patch; (2) A background noise glitch may cause the reading circuit to report the binary digit's incorrect value; and (3) The writing head may not report the incorrect value for the binary digit.

6.1.3 Digitalization of Information

The ingredients of a message are unrelated to its transmission: what the message expresses does not matter. It could be text, sound, picture, or video, but they are all 0's and 1's to the transmission channel. This digitalization principle unifies all information engineering, establishing that text, telephone signals, photographs, and the film could be encoded in bits for all communication forms. The underlying foundation of all we have is this digital representation.

6.1.4 Source Coding

Source code is any code written using a human-readable programming language, typically as plain text, with or without comments. Source code is also converted into binary machine code by an assembler or compiler that the program can execute. Source coding's fundamental goal is to eliminate redundancy in the information to make the message smaller (compressing the message content). The most frequently used for data compression is Huffman Coding today. Just some examples of its use are JPEGS, MP3s, and ZIP archives.

6.1.5 Entropy and Information Content

Information is an ordered symbol of sequences. The basic concept is to measure and transmit information in the form of entropy. The value associated with a single discrete random variable is defined as an information measure. It is measured as the opposite of the probability of something happening. For example, if an occurrence is highly likely, such as "the planet revolves around the sun" or "the sun rises from the east," the detail found in these phrases is similar to zero or rather small due to their exceedingly high probability of occurrence. While on the other hand, the information in the phrases "we will have a vaccine for the COVID-19 pandemic within this year" or "it will rain in Toronto today" is very high as the probability depends on a large number of factors, and the events are highly uncertain.

In communications and coding disciplines, information theory was initially formulated, and its implementation and use have now been extended to a wide variety of fields, including ML. Information theory concepts are used to define cost functions for optimization in parameter estimation problems, and information theory concepts are used to estimate unknown distributions of probability in the context of constrained optimization tasks.

The definition of data is part of our regular vocabulary, as is the case for probability. In this sense, either an event's occurrence is unknown to us or its occurrence probability is very remote; an incident carries information even though it occurs. Therefore, information is logically described as the negative logarithm of the event occurrence probability to formalize the notion of knowledge from a mathematical point of view. If the occurrence is likely to happen, the information's content is close to zero; if it is, however, an improbable event, the amount of information is extensive. For instance, if one tells us that the sun shines bright in the Sahara Desert during summer days, we might consider such a statement very dull and useless. On the contrary, suppose anyone gives us news during the summer about the snow in the Sahara: that comment carries many details. It can potentially spark a debate about climate change.

The information associated with any value $x \in X$ is denoted as $I(x)$ because of a discrete random variable, x, which takes values in the set X, and is defined as:

$$I(x) = -\log P(x) : \text{Information associated with } x = x \ X \qquad (6.1)$$

For the logarithm, any base may be used. Data are represented in terms of nats (natural units) if the natural logarithm is chosen. If the base 2 logarithm is employed, information is calculated using bits (binary digits).

It is also in line with common sense logic to use the logarithmic function to describe information that the information content of two statistically unrelated events should be the sum of the information communicated independently by each of them. Consider a binary event such as a coin flip where heads are symbolized by 1 and tails by 0 and assume that $P(1)=P(0)=0.5$. It is a binary random variable $x \in X = \{0, 1\}$. As a source that produces and emits two potential values, we may consider this random variable. The data content of both of the two equally likely occurrences is $I(0)=I(1)=-\log_2 0.5 = 1\text{bit}$.

The entropy in the context of information theory is viewed as the degree of uncertainty (or messiness) of a random variable that can take multiple states. It is the expected value of information from a discrete random variable's values and is also known to be the weighted information average for different outputs. The minimum estimated number of binary questions required to classify the random variable is directly related to it. More entropy means more questions to ask and more uncertainty about the identity of a random variable. The lower bound is given by the number of bits (units which vary) needed to encode symbols drawn from any P distribution. Entropy's lower or higher values depend on the distribution (indicating the probability that the result is definite or not).

From the concepts of conditional probability, we can derive the definition of mutual information. The information content generated by the occurrence of the event $y=y$ about the event $x=x$ is calculated by mutual information, denoted as $I(x; y)$ and defined by:

$$I(x, y) = \log \frac{P(x\,|\,y)}{P(x)}:$$ (6.2)

Provided the two discrete random variables, $x \in X$ and $y \in Y$. Note that their mutual knowledge is zero if the two variables are statistically independent; this is most fair because observing y says nothing about x. On the contrary, if it is inevitable that x will occur by observing y, as if $P(x|y) = 1$, the mutual knowledge will become $I(x, y)=I(x)$, which is again compatible with traditional reasoning that leads us to the expression:

$$I(x, y) = I(x) - I(x\,|\,y)$$ (6.3)

The source transmits binary symbols, x, to a communication medium, with a probability $P(0)=P(1)=1/2$. The channel is noisy, so the obtained symbols, y, may have changed polarity due to noise. For example, if we transmit data over a communication channel, i.e. a string of bits, there is some possibility that the message received will not be the same as the message transmitted. We would like to provide a communication medium for which this chance was nil or so near zero to be indistinguishable from zero for practical purposes. Transmitted bits are struck by noise, and the noisy (possibly wrong) data are what the receiver receives. Upon receipt of a series of symbols, which was the initially transmitted one, the receiver's task is to determine.

Consider a noisy disc drive that transmits each bit with probability $(1-p)$ correctly and with probability p incorrectly. This instance shows the effect of a communications channel in its simplest form. The purpose of our example is to evaluate the shared knowledge about the occurrence of $x=0$ and $x=1$ after observation of $y=0$ and vice versa. To this end, as in Figure 6.2, we first need to measure the marginal probabilities. Now let us consider $p=0$ and then $I(0, 0)=1$ bit, which is equal to $I(x=0)$, so with certainty, the output defines the input. If $p=0.5$, on the other hand, then $I(0, 0)=0$ bits, since the noise will change polarity at random with equal probability. If now $p=0.25$, then $I(0, 0)=\log_2 \dfrac{3}{2}= 0.587$ bits and $I(1, 0)=-1$ bit. Note that the mutual information can also take negative values. For the disc drive case, let us imagine that $p=0.1$, that is, the original and transmitted images are flipped by 10% of the bits on a disc drive, as seen in Figure 6.3. In its whole life span, a useful disc drive will flip no bits at all.

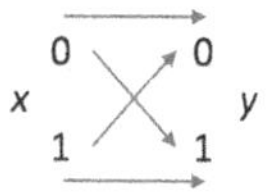

$$
x \quad \begin{matrix} 0 \\ 1 \end{matrix} \times \begin{matrix} 0 \\ 1 \end{matrix} \quad y \qquad \begin{cases} P(y = 0|x = 0) = 1 - p, & P(y = 0|x = 1) = p \\ P(y = 1|x = 0) = p, & P(y = 1|x = 1) = 1 - p \end{cases}
$$

FIGURE 6.2 The binary symmetric channel. The symbol transmitted is x, and the symbol obtained is y. The degree of noise, the likelihood of a bit being flipped, is p.

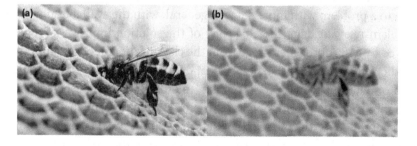

FIGURE 6.3 A binary data sequence of length 10,000 transmitted over a binary symmetric channel (a) original image, (b) with noise level $p=0.1$.

6.2 DATA HANDLING

Information is extracted from data collected from almost every aspect of life, stored, transmitted, transformed, and manipulated to make sense of the information it contains. With computers' assistance, the traditional approach to solving issues is to write programs that solve the problem. The programmer must comprehend this method's problem, find a suitable solution, and enforce it on the computer. Since the person deduces the solution from the problem formulation, we call this method deductive.

However, a large amount of data is generated in physics, chemistry, biology, medicine, and other fields of sciences, which is difficult for people to understand and interpret. A computer that learns can also find a solution to a problem. Such a computer analyzes the data and finds structures in the data automatically, i.e. learns. It is possible to use information about the extracted structure to solve the problem at hand. ML is about inductively solving problems by machines, i.e. computers. We call this method inductive.

ML aims to extract from the data the relevant information and make it accessible to the user. What are we talking about with "relevant information"? We typically have a specific question in mind when analyzing data, such as "how many types of cars can be identified in this video" or "what will be the weather next week." So, the answer can take the form of a single number (there are five cars, or a number series, or next week's temperature) or a complicated pattern (next week's cloud configuration). We like to visualize it using graphs, bar plots, or even small movies if our question's response is complicated. However, one must keep in mind that the exact analysis depends on the assignment in question. To achieve the objective in mind, one must perform quite a few tasks in ML.

ML researchers create algorithms that automatically enhance a solution to a problem with more data. In general, with the available amount of problem-relevant data, the efficiency of the solution improves. Problems solved by ML approaches range from the classification of observations, the prediction of values, the structuring of data (e.g. clustering), the compression of data, the visualization of data, the filtering of data, the collection of appropriate data components, the extraction of dependencies between data components, the modeling of data generation systems, the creation of noise models for observed data, and the incorporation of data from various sensors. Let us discuss some of the critical tasks that are typically performed in ML.

6.2.1 Supervised Learning

If a computer is supplied with labeled training data (the class) by a human expert, the setup is known as supervised learning. Here a human expert acts as a teacher to provide the correct answer to the machine, and the machine tries to reproduce the answer. The calculation is carried out for every single object. This calculation implies that on each object, the model provides an error value. The problem can generally be presented as finding (or learning) a function $f(x)$ that approximates any input x to the correct class labels. For example, we can decide that our class label's predictor is $sine[f(x)]$.

Often, noise corrupts the vectors in the training set. Two kinds of noise are available. The noise class randomly alters the function's value; the noise attribute randomly alters the input vector components' values. Any case requiring that the hypothesized feature corresponds precisely with the samples' values in the training set would be unacceptable. The induced function is typically evaluated in supervised learning on a different set of inputs and function values called the test set for them. When it guesses well on the testing range, a hypothesized feature is said to generalize. Suitable metrics are both mean-squared error and the total number of errors.

Typical fields of supervised learning are classification, regression (assigning a real value to the data), or time series analysis (predicting the future).

6.2.1.1 Classification

In a wide variety of human activities, the process of classification exists. In its broadest sense, the word may encompass any context in which a decision or a prediction is taken based on the knowledge currently available. A classification procedure is then a systematic process for making such decisions repeatedly in new contexts. It has also been named pattern recognition or discrimination to create a classification procedure from a collection of data for which the proper classes are identified. The following are three examples of contexts in which a classification task is required: (1) mechanical processes for sorting letters based on machine-read postcodes; (2) awarding credit status to persons based on financial and other personal information; and (3) preliminary diagnosis of a patient's condition to select urgent treatment while pending definitive test results.

In classification, the aim is to assign an unknown pattern to several classes considered to be known. For example, in X-ray mammography, we are given an image where an area implies a tumor's presence.

A computer-aided diagnostic system's objective is to predict whether this tumor belongs to the benign or malignant category. In reality, using complex and sometimes very comprehensive data, some of the most urgent problems emerging in research, industry, and commerce can be considered classification or decision problems.

6.2.1.2 Regression

The regression shares the generation/selection stage of the function to a large degree. However, the output variable, y, is not inherently discrete but takes values in the real axis interval or a region in the plane of complex numbers. The regression task is a curve-fitting problem. A collection of training points is given to us, and the task is to estimate a function f, whose graph matches the details. Once we have found such a function, we can predict its output value when an unknown point arrives.

For example, given current market conditions and all other relevant details in financial applications, one can predict tomorrow's stock market price. A calculated value of a corresponding function is any piece of knowledge. Signal and image restoration tasks come under this prominent umbrella of tasks. It is also possible to see signal and picture denoising as a unique form of a regression task.

6.2.1.3 Time Series Prediction

Future values must be estimated based on current and past values in a time series prediction task. For example, if we track the length, width, and weight of the fish every day or every week from birth and predict its height, weight, or health status as a grown fish, it will be a prediction activity. Suitable fish will be selected early if such predictions are accurate.

6.2.2 Semisupervised Learning

There might be cases where some labeled data and some unlabeled data are available. One would prefer to use the labeled data to learn a classifier, but tossing out all that unlabeled data seems wasteful. The main question is: What can one do to help learning with that unlabeled data? Furthermore, for this to be beneficial, what assumptions do we have to make? One suggestion is to try to use the unlabeled data to discover a better judgment cap. One can accomplish this in a discriminatory environment by finding decision limits that do not travel too closely to unlabeled results. One may treat some of the labels as observed and some as concealed in a generative sense. This type of learning is called semisupervised learning.

6.2.3 Unsupervised Learning

If ML methods are processing a collection of unlabeled data and automatically clustering the data into smaller sets containing related objects, it is an example of unsupervised learning. The extraction of data structures also leads to the discovery of definitions, subjects, abstractions, variables, triggers, and more, all of which mean the same thing. These are the underlying semantic variables that can explain the details. Knowing these variables is like denoising the information. The uninteresting bits and pieces of the signal are first stripped off and then converted into an often-lower dimensional space that reveals the underlying variables.

The consistency of models on unsupervised problems is often calculated by the cumulative output on all objects instead of supervised problems. Unsupervised learning measurements usually include the following: the content of information, the orthogonality of the components developed, the statistical independence, the variance explained by the model, the likelihood that the data observed can be produced by the model (later implemented as a probability), and distances between and within clusters. Projection methods, clustering methods, density estimation, or generative models are common areas of unsupervised learning. Unsupervised approaches seek to extract the data structure, represent the data in a more compact or more useful way, or create a data generation process model or sections of it.

6.2.3.1 Projection

Visualizing it is a straightforward thing that one might want to do with data. Unfortunately, visualizing items in more than two or three dimensions is complicated because much of the data are in hundreds of dimensions (or more). The problem of taking high-dimensional data and embedding it in a lower-dimensional space is the reduction of dimensionality. Methods of projection create a new representation of objects as a function vector given a representation of them. In most cases, to eliminate redundancies and components that are not necessary, they downproject feature object vectors into a lower-dimensional space. Principal component analysis (PCA) represents the object in some orthogonal directions via feature vectors, giving the data extension. The directions are ordered so that the first direction gives the maximum data variance direction, the second direction gives the orthogonal maximum data variance to the first component, and so on. Independent component analysis goes a step further than PCA and describes the artifacts by statistically mutually independent function components. By adding Gaussian noise to each original component, factor Analysis extends PCA and assumes the components' Gaussian distribution. Therefore, projection pursuit lookups for non-Gaussian components may contain exciting details.

6.2.3.2 Clustering

Methods of clustering are searching for data clusters and therefore finding structure in the data. Putting a collection of objects into classes that are identical to each other is one way of expressing regularity. For instance, biologists have observed that most artifacts fall into two groups in the natural world: brown things run away, and green things do not run away. They name the first party animals and the second plants. This operation of grouping items together is called clustering. SOMs (self-organizing maps) are a type of clustering system that also performs a downprojection in order to visualize the data. The downprojection holds clusters in the vicinity. The clustering methods or (down)projection can be viewed as feature construction methods since the new components can now represent the object. The clustering object includes the cluster to which it is nearest or an entire vector describing the distances between the various clusters.

6.2.3.3 Density Estimation

Methods of density estimation aim to generate the density from which the data have been collected. Density estimation is constructing an approximation of an unobservable underlying probability density function based on observed data. The unnoticeable density function is assumed to be the density by which a large population is distributed; the data are generally considered a random sample. Numerous density estimation methods are used, with the most basic density estimation being a rescaled histogram. Learning problems are characterized by any unspecified probability distribution D over input/output pairs (x, y). Suppose one is aware of what D is. Mainly one is provided with a function that took two inputs, x and y, and returned the probability of the x and y pair under distribution D. When one has access to such a function; classification becomes simple. For any test input x, we can define the density estimate as to the classifier that returns the density estimate. $\hat{y}$ maximizes the calculated values $(\hat{x}, \hat{y})$.

6.2.3.4 Generative Models

Modeling the data generation method is another ML approach to understand the data collection process. These are called generative models that aim to construct a model that describes the density of the data observed compared with density estimation methods. The objective is to obtain a world model for which the model's data point density coincides with the data density observed. Models that generate the distribution observed for real-world data are chosen, so these models explain or reflect the process of data generation. The process of data generation may also have input components or random components that drive the process. These inputs or random parts can be used in the model. The generative approach needs to have as much prior knowledge about the environment or desired model properties in the model to reduce the number of models explaining the observed data. A generative model can forecast data generation for unnoticed inputs, predict the data generation process's behavior if its parameters are modified externally, generate data from artificial training, or predict unexpected events. In particular, modeling methods may provide new insights into the functioning of complex world structures, such as the brain or the cells.

6.3 LEARNING INPUT AND OUTPUT FUNCTIONS

We are using Figure 6.4 to describe the problem of learning a function with some of the terminologies used. Imagine that a function, f, exists, and the learner's task is to guess what it is. Our assumption about the function to be learned is denoted by h. Both f and h are input vector-valued functions $X=(x_1; x_2... x_i...x_n)$ that have n components. We believe that h is implemented by a device with X as its input and $h(X)$ as its output. Both f and h may be vector-valued themselves. We assume that the hypothesized function, h, is selected from the function class H. We sometimes know that f belongs to this class or a subset of this class as well. Based on a training set, ξ, we select h from m input vector examples. Many critical facts influence the nature of the conclusions taken regarding all of these entities.

Two types of learning modes, supervised and unsupervised, have already been defined. We know (sometimes only approximately) the values of f for the m samples in the training set, ξ, in supervised learning.

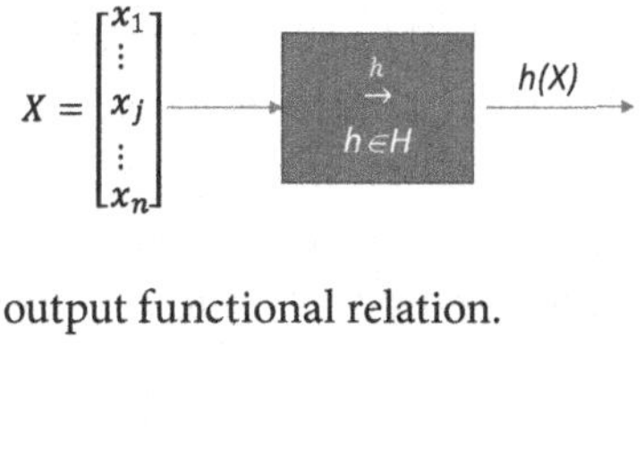

FIGURE 6.4　Input and output functional relation.

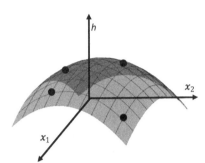

FIGURE 6.5 A surface that can be used to fit four data points.

We assume that if we can find a hypothesis, h, which for the members of ξ closely agrees with f, this hypothesis will be a good guess for f – especially if ξ is large. A simple instance of supervised learning of a function is curve-fitting. Suppose at the four sample points shown by the solid circles in Figure 6.5; we are given the values of a two-dimensional function, f. We want these four points with a function, h, drawn from the second-degree function set, H. Over the x_1, x_2 plane, we show a two-dimensional parabolic surface that fits the points. The function, f, that generated the four samples is this parabolic function, h is our hypothesis. For the four samples in this situation, $h = f$, but we do not need specific matches.

We have a training set of vectors without functional values for them in the context of unsupervised learning. Usually, the issue, in this case, is to partition the training set into subsets, $\xi_1 \ldots \xi_R$, in any suitable manner. The problem can still be assumed to be one of learning a function; the function's value is the subset's name to which an input vector belongs. In taxonomic problems, unsupervised learning methods are used to invent ways to classify data into meaningful categories.

6.3.1 Input Vectors

Since methods of ML originate from many different traditions, synonyms are rife in its terminology. For example, several names are used to call the input vector. Some of these are vector input, vector pattern, vector function, sample, instance, and example. Attributes, input variables, and components are variously called the components, x_i, of the input vector.

The component values may consist of three fundamental types. They may be numbers with real values, numbers with discrete values, or categorical values. As an example of categorical values, knowledge about a student may be represented by the values of the class, major, sex, and supervisor attributes. A single student will then be represented by such a vector (sophomore, history, male, Higgins). Additionally, categorical values (small, medium, large) or unordered values can be ordered (as in the example given). Mixtures of all these types of values are also, of course, possible.

It is possible to represent the input in the unordered form in all cases by listing the attribute names along with their values. The vector form assumes that a form orders the attributes and implicitly gives them. As an example of representing an attribute value, we may have the following: major: history, sex: male, class: sophomore, adviser: Higgins, age: 20. The vector form will be used exclusively by us. Boolean values are used in an essential specialization, which can be viewed as a particular case of either discrete numbers (1,0) or categorical variables (true, false).

6.3.2 Output

The output may be a real number, in which case a function estimator is called the mechanism embodying the function, h, and the output is called an output value or estimate. Alternatively, the output can be a categorical value. A classifier, a recognizer, or a categorizer is called the mechanism embodying h, and the output itself is called a name, a class, a category, or a judgment.

In various recognition problems, classifiers are used, such as in recognition of hand-printed characters. In that case, the input is an appropriate representation of the printed character, and the classifier maps this input into one of, say, 64 categories. With sections being actual numbers or categorical values, vector-valued outputs are also possible. That of Boolean output values is a particularly critical case. A training pattern with value 1 is considered a positive example in that case, and a training sample with value 0 is called a negative example.

The classifier implements a Boolean function when the input is Boolean, too. In some depth, we research the Boolean case because it enables us to make general critical points in a simpler environment. Sometimes, learning a Boolean function is called concept learning, and the function is called a concept.

6.4 BAYESIAN DECISION THEORY

This section will look at the inference of training data as a random process modeled by the joint distribution of probability over variables of input (measurements) and output (say class labels). Estimating the underlying distribution is a difficult process in general. However, there are several constraints or "trade tricks," so to speak, that make this task manageable and useful under certain conditions. Sampling theorists focus, given minimal assumptions, on making methods guaranteed to work most of the time. Given the unique data collection, the Bayesians try to make inferences that consider all available information and address the question of interest. The sampling theory is the commonly used approach in statistics, and most research publications disclose their studies using quantities such as confidence intervals, levels of significance, and p-values.

A p-value (e.g. $p=0.05$) is the probability that the outcome will be as extreme as or more extreme than the observed outcome, given a null hypothesis for the probability distribution of the results. Untrained readers generally interpret such a p-value as if it were a Bayesian probability (for example, the posterior probability of the null hypothesis), and perhaps more worryingly, an interpretation that both sampling theorists and Bayesians would accept is wrong. While the responses from a Bayesian approach and sampling theory are very similar, we can also find instances where substantial differences exist. Sampling theory can be trigger-happy in some situations, arguing that the results are so impossible that "the null hypothesis can be ignored." In contrast, those results are weakly confirmed by the null hypothesis. As we can now see, there are also inference concerns, where sampling theory fails to recognize "important evidence," but a Bayesian approach and everyday intuition agree that the evidence is solid. Most telling of all are the inference problems where, depending on irrelevant factors concerned with the design of the experiment, the "significance" assigned by sampling theory varies.

We will assume a discrete universe to make it straightforward, i.e. values our random variables take on are finite numbers. For example, consider two random variables X that takes on possible k values $x_1, x_2, ...,$ x_k, and H with two values $h_1; h_2$. X values could stand for a person's body mass index (BMI) measurement (weight over height), and H stands for the two possibilities, $h1$ standing for the "person being overweight" and h_2 as the "person of normal weight" possibility. We want to estimate the likelihood of the person being overweight given a BMI measurement. A two-dimensional array (two-way array) with 2k entries (cells) is the joint probability $P(X, H)$. Every example of training (x_i, h_j) falls into one of those cells, so

$P(X = x_i, H = h_j) = P(x_i, h_j)$ holds the ratio between the number of cell hits (i, j) and the total number of examples of training. Consequently,

$$\sum_{ij} P(x_i, h_j) = 1 \qquad (6.4)$$

By summing over columns or rows, the projections of the array on its vertical and horizontal axes are called marginalization and generate:

$$P(h_j) = \sum_i P(x_i, h_j) \qquad (6.5)$$

The sum over the jth row is the probability $P(H = h_j)$, i.e. the probability of an individual being overweight (or not) before any calculation is seen – these are called priors. Similarly:

$$P(x_i) = \sum_j P(x_i, h_j) \qquad (6.6)$$

This expression is the probability that $P(X = x_i)$ is the probability of obtaining such a BMI measurement, which is sometimes referred to as proof. Notice that $\sum_j P(h_j) = \sum_i P(x_i) = 1$ by definition. The conditional probability $P(h_j|x_i) = P(x_i, h_j) / P(x_i)$ is the ratio between the number of hits in the cell (i, j) and the number of hits in the ith column, i.e. the probability that, given the measurement $X = x_i$, the outcome is $H = h_j$. This expression gives rise to the Bayes formula:

$$P(h_j|x_i) = \frac{P(x_i|h_j)P(h_j)}{P(x_i)} \qquad (6.7)$$

$P(h_j|x_i)$ is called the posterior probability on the left-hand side, and $P(x_i|h_j)$ is called the conditional probability class. The Bayes formula provides a way for the prior, proof, and class probability to estimate the subsequent probability. It is useful when the class probability is normal to measure (or collect data from), but it is not very convenient to calculate the posterior directly.

To understand the Bayesian inference and sampling theory's applicability, let us consider the race to develop an effective vaccine for the COVID-19 pandemic. Two vaccines A and B are tested on a group of 40 volunteers. Vaccine B is a control treatment with no active ingredients. Out of the 40 volunteers, 30 are randomly selected for treatment A, and the remaining 10 are given treatment B. The subjects are then observed for a certain period after their vaccination. Of the 30 in group A, 1 in the observation period contracted COVID-19, while 10 in group B, 3 contracted COVID-19. Is treatment A superior to treatment B?

6.4.1 Sampling Theory Approach

The standard approach of sampling theory to the question "is A better than B?" is constructing a statistical test. Typically, the test contrasts a hypothesis like H_1: "A and B have different efficacy" with a null hypothesis like H_0: "A and B have the same efficacy." A novice might object, "no, no"; "A is better than B" hypothesis with the "B is better than A" alternative should be tested! Nevertheless, in sampling theory, such objections are not welcome. The first hypothesis is barely discussed again until the null hypotheses have been defeated – attention focuses entirely on the null hypothesis. The null hypothesis is solely accepted or rejected based on how surprising the data was to H_0, not how much better H1 anticipated the data. One selects a metric that tests how far from the null hypothesis a data set deviates. In this case, χ^2 (chi-squared) will be the typical statistic to use.

To calculate χ^2, assuming the null hypothesis is true, we take the difference between each data measurement and its expected value and divide the square of that difference by the variance of the measurement, assuming the null hypothesis is true. The four data measures in the present problem are the integers FA+, FA−, FB+, and FB−, i.e. the number of subjects receiving treatment A who contracted COVID-19 (FA+), the number of subjects receiving treatment A who did not develop COVID-19 (FA−). The description of χ^2 is:

$$\chi^2 = \sum_i \frac{\left(F_i - F_i\right)^2}{F_i} \tag{6.8}$$

Any simple book of statistics would suggest the application of Yates' correction, which gives:

$$\chi^2 = \sum_i \frac{\left(\left|F_i - \langle F_i \rangle\right| - 0.5\right)^2}{\langle F_i \rangle} \tag{6.9}$$

In this case, the expected counts are $F_{A+} = f_+ N_A$, $F_{A-} = f_- N_A$, $F_{B+} = f_+ N_B$, $F_{B-} = f_- N_B$, provided the null hypothesis that treatments A and B are equally efficient and have rates f_+ and f_- for the two results.

Based on how large χ^2 is, the test supports or rejects the null hypothesis. We have to determine the sampling distribution of χ^2 to make this test reliable and give it a "significance level." Considering that the four data points are not independent (they follow the two constraints $F_{A+} + F_{A-} = N_A$ and $F_{B+} + F_{B-} = N_B$) and the parameters $f_{\pm}$ are not defined. The number of degrees of freedom in the data is reduced from four to one by these three constraints. When the information arrives, sampling theorists estimate the unknown parameters from the data for the null hypothesis and test χ^2. Let us choose 5% as our level of significance. With one degree of freedom, the critical value for χ^2 is $\chi^2_{0.05} = 3.84$.

Furthermore, χ^2, as obtained from Equation 6.8, is $\chi^2 = 5.93$. Since this value exceeds 3.84, at the 0.05 significance level, we reject the null hypothesis that the two treatments are equivalent. Note that this response does not say how much more productive A is than B. It merely states that A is "significantly" different from B. Furthermore, here, "significant" means just "statistically important" and not functionally significant. However, we find $\chi^2 = 3.33$ if we use Yates's correction (Equation 6.9) to support the null hypothesis. This finding has little to do with what we are looking for, which is how likely it is that treatment A is superior to treatment B.

6.4.2 Bayesian Inference Approach

Let us find out what we are looking for right now. We discard the hypothesis that the two treatments are equally effective because we do not believe it. The probabilities of developing the disease by individuals, provided therapies A and B, respectively, are two uncertain parameters, pA+, and pB+. We can infer these two probabilities, provided the results, and by analyzing the posterior distribution, we can address questions of interest. The posterior distribution is:

$$P\left(p_{A+}, p_{B+} | \{F_i\}\right) = \frac{P\left(\{F_i\} | p_{A+}, p_{B+}\right) P\left(p_{A+}, p_{B+}\right)}{P\left(\{F_i\}\right)} \qquad (6.10)$$

The probability function is:

$$P\left(p_{A+}, p_{B+} | \{F_i\}\right) = \binom{N_A}{F_{A+}} \left(p_{A+}\right)^{F_{A+}} \left(p_{A-}\right)^{F_{A-}} \binom{N_B}{F_{B+}} \left(p_{B+}\right)^{F_{B+}} \left(p_{B-}\right)^{F_{B-}}$$

$$= \binom{30}{1} \left(p_{A+}\right)^1 \left(p_{A-}\right)^{29} \binom{10}{3} \left(p_{B+}\right)^3 \left(p_{B-}\right)^7 \qquad (6.11)$$

What prior distribution do we use? The prior distribution allows us to use information from other experiments or a previous assumption that the two p_{A+} and p_{B+} parameters have similar values. However, they are different from each other. The most straightforward prior distribution, a uniform distribution over each parameter, will be used here.

$$P\,(p_{A+},p_{B+})=1 \tag{6.12}$$

Assuming the prior distributions p_{A+} and p_{B+} are separable, the posterior distributions also become separable.

$$P\left(p_{A+},p_{B+}|\{F_i\}\right)=P\left(p_{A+}|F_{A+},F_{A-}\right)P\left(p_{B+}|F_{B+},F_{B-}\right) \tag{6.13}$$

If we want to understand the answer to the issue, how likely is p_{A+} smaller than p_{B+}? We can answer precisely that question by calculating the posterior probability.

$P(p_{A+} < p_{B+}|Data)$, which is the integral of the posterior joint probability $P\left(p_{A+},\,p_{B+}|Data\right)$ over the region in which $p_{A+} < p_{B+}$; the value of this integral (acquired by a simple numerical integration of the probability function (6.11) over the corresponding region) is:

$$P(p_{A+} < p_{B+}|Data)=0.99 \tag{6.14}$$

We believe there is a 99 percent probability that treatment A is preferable to treatment B based on the facts and our previous conclusions. Finally, data (1 out of 30 after vaccination A contracted the disease, and 3 out of 10 after vaccination B contracted the disease) offer unequivocal proof (about 99 to 1) that treatment A is superior to treatment B, according to our Bayesian model. It is also easy to answer other related questions under the Bayesian approach. If we want to know, for instance, how likely is treatment A to be ten times more successful than treatment B? The joint posterior probability $P\left(p_{A+},\,p_{B+}|Data\right)$ can be integrated over the area in which $p_{A+} < 10\;p_{B+}$.

6.5 NEURAL NETWORKS

The field of neural networks has originated from numerous sources, ranging from humanity's obsession with understanding and emulating the human brain. This desire extends to broader issues of copying human skills such as expression and language used to the realistic disciplines of pattern recognition, modeling, and prediction in the commercial, science, and engineering fields. There are about 60–100 billion neurons in the brain. Via elementary structural and functional units/links, known as synapses, each neuron is linked with other neurons. There are between 50 and 100 trillion synapses in the brain. These connections mediate information between associated neurons. Chemical pulses that convert electrical pulses produced by a neuron to a chemical signal and then back to an electrical one are the most common type of synapses. A synapse is either triggered or inhibited, depending on the input pulse(s).

Each neuron is linked to other neurons through these connections, and this occurs in a layer-wise fashion in a hierarchically ordered manner. A breakthrough from the learning theory perspective occurred in 1943 when a primary neuron computational model was created to connect neurophysiology with mathematical logic. This model showed that given enough neurons and the necessary adjustment of synaptic ties, each represented by weight, any computational function could, in principle, be computed. Neural networks are learning machines that are linked in a layered fashion, comprising many neurons. To minimize a preselected cost function, learning is accomplished by changing the unknown synaptic weights.

In the perceptron, we view the input data point (e.g. an image) as directly linked to an output (e.g. label). Since there is only one layer of weight, this is also called a single-layer network. Instead of directly connecting the inputs to the outputs, we can insert a layer of "hidden nodes," switching from a single-layer network to a multilayer network.

One of the critical drawbacks of linear models, such as perceptron and linear regularized models, is that they are linear! In other words, they are unable to grasp the limits of unreasonable decisions. Decision trees and k-nearest neighbor (KNN), on the other hand, can learn arbitrarily complex decision boundaries. One method is to chain a series of perceptrons together to create more intricate neural networks. Figure 6.6 shows an example of a two-layer network. Here, five inputs (features) that are fed into two hidden units can be seen. Then these hidden units are fed into one single unit of output. A different weight corresponds to each edge in this diagram. (Even though it looks like there are three layers, this is called a two-layer network because we do not consider the inputs as a real layer. That is, it is two layers of trained weights.)

Using a neural network prediction is a straightforward generalization of prediction with a perceptron. Second, based on the inputs and the input weights, one measures node activations in the hidden layer. Then, given the concealed unit activations and the second layer of weights, one computes

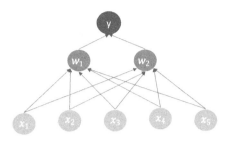

FIGURE 6.6   A two-layer neural network with five inputs and two hidden units.

the output unit activations. The only significant difference between the perceptron computation and this computation is that the hidden units compute their inputs' nonlinear function. Typically, this is referred to as the activation function or connection function. More formally, if $w_{i,\,d}$ is the weights on edge connecting input "d" to remote unit "i," then hidden unit "i" activation is calculated as $h_i = f(w_i \cdot x)$, where f is the connection function, and w_i refers to the weight vector feeding into the "i" node. The hyperbolic tangent function, tanh, is a standard connection function. Assuming we use tanh as the connection function, it is possible to compute the general prediction made by a two-layer network as follows. The function adopts a weight matrix W corresponding to the first layer's weights and a weight vector v corresponding to the second layer's weights. This entire calculation can be written in one line as:

$$\hat{y} = \sum_i v_i \tanh\left(w_i \cdot \hat{x}\right) = v \cdot \tanh\left(W\hat{x}\right) \tag{6.15}$$

The second line here is shorthand if hyperbolic tangent function can take a vector as an input and a vector as an output product.

The point is that there are more expressive two-layer neural networks than single-layer networks (i.e. perceptron). The key question when dealing with two-layer networks is how many hidden units one should have? The total number of parameters is $(D+2)$ K if your data are D dimensional and you have K hidden units. (First +1 is from the bias, the second is from the second weight layer.) Following the heuristic that for each parameter one is trying to estimate, one can have one to two instances. This estimate indicates a formula for choosing the number of hidden units as roughly N/D. In other words, one can safely have plenty of hidden units if one has loads and loads of examples. One should probably limit the number of hidden units in one's network if one only has a few examples. The number of units is both a form of inductive bias and a form of regularization. The number of hidden units in both views controls how complex one's feature would be. Lots of hidden units suggest very complex features. As the number of hidden units increases, training performance can get better, but since the network starts to overfit the data, test performance gets worse at that point.

6.5.1 Multilayer Neural Networks

We understand from the discussion above that two-layer networks are universal approximators of functions. This understanding raises the question: Why do we think about deeper (or multilayered) networks if two-layer networks are so great? We may borrow some ideas from computer science theory, namely the concept of circuit complexity, to understand the response. The aim is to demonstrate that there are functions for which using a deep network could be a "good idea." In other words, if one pushes the network to be shallow, some functions will require a large number of hidden units but can be accomplished in a limited number of units if one allows it to be deep.

A different definition of the input patterns is given for each layer of the neural network. The input layer represents each pattern as a point in the feature space. The first hidden layer of nodes forms a partition of the input space and uses a coding scheme of zeros and ones to position the input point in one of the regions as outputs of the respective neurons. This partition can be viewed as a more abstract reflection of our patterns of data. Based on the information given by the previous layer, the second hidden layer of nodes encodes class-specific information; this is an additional representation abstraction that carries some "semantic meaning." For example, it provides information on whether a tumor is malignant or benign in a related medical application.

The hierarchical form of representation of the input patterns mimics how a human brain follows the world around us to "understand" and "sense." For example, in our visual system, this hierarchy includes the identification of edges, and primitive shapes initially that lead to more complex visual shapes formation as we progress to higher hierarchy levels before a semantic definition is eventually developed. Our brain is a multilayer architecture of 5–10 layers dedicated only to our visual system in this context.

An equivalent input–output representation can be obtained through a relatively simple functional formulation (such as the one indicated by the support vector machines) or through networks with fewer than three layers of neurons/processing elements. This representation is achieved perhaps at the cost of more elements per layer, which poses a problem.

The response is yes if the relationship of input–output dependency is easy enough. For more complicated activities, however, where more complex concepts need to be taught, such as recognizing a scene in a video clip and recognizing language and voice, the underlying functional dependency is very complex so that we cannot express it analytically. We consider representational compactness meaning if it consists of relatively few free parameters (few computational elements) to be learned/tuned during the training process, a network, understanding an input–output functional dependency, is compact. Thus, we expect compact representations to result in better generalization results for a given number of training points. It turns out that one can obtain more compact representations of the input–output relationship using networks with more layers.

Even though such an argument for general learning tasks cannot be proven through theoretical findings, the Boolean function circuit theory's theoretical results suggest the following. Suppose a function can be realized compactly by, say, k layers of logical elements; it can require an exponentially large number of elements for $k-1$ layers to realize it. These arguments can seem a little misleading since we have already mentioned that two-layer node networks are universal approximators for a specific function class. This argument does not reveal, however, how one would accomplish this in reality. Any continuous function, for instance, can be approximated arbitrarily by a sum of monomials. Nevertheless, it may need many monomials, which is not technically feasible. We must be concerned with what is feasibly "learnable" in each representation in any learning assignment.

A block diagram of a deep neural network with three hidden layers is shown in Figure 6.7. The random input vector is denoted as "x," and those associated with the hidden variables are denoted as h_i, $i = 1, 2, 3$, and "y" is the output node vector. Pretraining progresses sequentially, beginning with the weights that link the input nodes to the first hidden layer nodes. This linking is done by optimizing the probability of the input observation samples observed, x, and treating the first layer's variables as hidden ones. Once the weights corresponding to the first layer have been determined, the respective nodes' output value can fire, and a vector of values is generated, h_1. This is the reason that leads us to adopt a generative model for unsupervised pretraining to produce outputs at the hidden nodes in

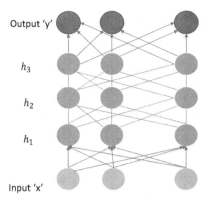

FIGURE 6.7   A deep neural network with three hidden layers.

a probabilistic way. These values are used in turn as observations for the next hidden layer's pretraining, and so on.

After some intensive study duration, multilayer perceptrons lost their original glory. They were largely superseded by other techniques, such as kernel-based schemes, support vector machines, and Bayesian learning methods. A significant explanation for this loss of popularity was that their training could become challenging, and algorithms related to back-propagation are often trapped in local minima. Although improvements may be accomplished by attempting various practical "tricks," such as multiple training using random initialization, their generalization performance may still not be competitive with other methods. If more than two hidden layers are used, this downside becomes more serious. The more layers one uses, the more challenging the training becomes, and the likelihood is increased to recover solutions that lead to low local minima.

An active field of research is seeking good ways of training deep networks. Two general strategies exist. The first is to try, mostly via a layerwise initialization technique, to initialize the weights better. It is always possible to do this using unlabeled data. Following this initialization, for any classification problem, one is interested in, backpropagation can be used to fine-tune the weights. Instead of gradient descent, a second approach is to use a more complicated optimization technique.

6.6 SUPPORT VECTOR MACHINES

Support vector machines are the highest performing methods in different scientific domains. Let us consider our original objective of linear classifiers: Find a hyperplane that distinguishes the positive examples of

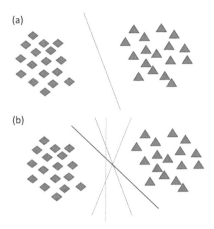

FIGURE 6.8 (a) A linearly separable problem where Class 1 data points (diamonds) are located on one side of the hyperplane (a two-dimensional line) and Class 2 data points (triangles) are located on the other side of the hyperplane. (b) Different solutions for dividing groups linearly.

training from the negative ones. Linear separability implies a discriminating function for the training data that describes a classification function. All positive examples are on one side of the boundary function, and the other side is all negative examples. For a linearly separable problem, Figure 6.8a demonstrates a two-dimensional example. A training set of instances $x_i \in R$,

$i = 1, 2, \ldots, n$, and class labels $y_i = \pm 1'$ are given (i.e. the training set is made up of positive and negative examples). We want to find a hyperplane trajectory $w \in R$ and a scalar b offset such that $w \cdot x_i + b > 0$ for positive examples and $w \cdot x_i + b < 0$ for negative examples. However, the data can be separated by various hyperplanes, as shown in Figure 6.8b for a two-dimensional example. Which is the hyperplane with the best separation?

There are infinitely many possible hyperplanes that can classify the training data. We need to add constraints to discover the best hyperplane that helps us find the most sensible solution. Another issue is that the structure is very constrained because it is improbable that a linear separation function will exist, to start with, for most real-world classification problems. Therefore, at a fair cost, we need to find a way to expand the system to include nonlinear decision boundaries. As for the first problem, since more than one separation hyperplane (assuming the training data are linearly separable), we need to ask ourselves: Which one of all those solutions has the best attributes for generalization?

Our aim is not generally to do very well (or correctly) on the training data in building a learning machine since the training data are simply a sample of the example space and not an accurate representation – it is merely a sample. Therefore, doing well on the specimen (training data) does not necessarily ensure (or even imply) that we will do well on the entire instance space. Building a learning machine aims to optimize the performance on the test data (the cases we have not seen), which means that we want to generalize good performance of classification on the training set over the entire instance space.

Consider a subset of all hyperplanes having a fixed margin, where the margin is defined as the distance to the hyperplane from the nearest training data point. We seek a separate hyperplane in the support vector machine that simultaneously minimizes the empirical error and maximizes the margin. The concept of optimizing the margin is intuitively appealing because it is less likely that a decision boundary close to any of the training instances would generalize well. After all, the learning system will be prone to tiny perturbations of such sample vectors. Figure 6.9 offers an understanding of why a greater margin is ideal because data points can be noisy without jumping over the boundary function. Our theoretical factors recommend using the function of classification with the lowest complexity. We will use the margin to calculate the lowest complexity. Mathematically, it is written as a constrained optimization problem:

$$\min_{w,b} \frac{1}{2} w \cdot w, \text{ such that } y_i (w \cdot x_i + b) - 1 \geq 0, i = 1, 2, \ldots, n \qquad (6.16)$$

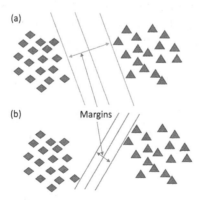

FIGURE 6.9 Intuitively, better generalization on the top pair of the data set (a) is predicted from separation than from the bottom pair (b) since greater distance to the data points allows noisy data to be better cushioned.

The formulation above is in primal space. However, we can solve this problem in dual space if we use Lagrange multipliers. The Lagrange function is:

$$L(w,b,\mu)=\frac{1}{2}w\cdot w-\sum_{i=1}^{n}\mu_i\left[y_i(w\cdot x_i+b)\right]-1 \qquad (6.17)$$

where $\mu_i \geq 0$ are Lagrange multipliers; multipliers are more significant than zero since the constraints are inequalities. The solution to the problem of optimization is a Lagrangian saddle point. The minimization must be performed over w and b and the maximization over $\mu_i \geq 0$ to find the saddle point. However, this optimization problem needs the training data to be linearly separable – a requirement that may be impractical. By introducing the definition of a "soft margin" in which the separability holds roughly with some error, we can relax this condition.

Consider now the more practical case of overlapping groups with "soft margins" and the related geometric representation of the role of classification. There is no linear classifier that can correctly identify all the points in this situation, and some errors are bound to occur. For a linear classifier, Figure 6.10 displays the respective geometry. There are three types of points, points on the boundary or beyond the margin on the classifier's correct side, and these points do not commit margin error (Figure 6.10a). Points that are on the right side of the classifier but within the margin (Figure 6.10b) cause a margin error greater than 0 but less than 1. Points that are on the wrong side of the classifier (Figure 6.10c) commit an error greater than or equal to 1. Our efforts would be to identify a hyperplane classifier that optimizes the margin and at the same time keeps the number of errors (including margin errors) as small as possible.

FIGURE 6.10 There are three types of points where groups are overlapping: (a) points that lie outside or on the boundaries of the margin and are correctly classified ($\xi_n=0$); (b) points inside the margin and correctly classified ($0<\xi_n<1$) denoted by red symbols; and (c) misclassified points denoted by a red symbol ($\xi_n \geq 1$).

The use of slack parameters is the key idea. The following is the intuition behind slack parameters. Suppose that we find a set of w, b parameters that do an excellent job on 9,999 data points. The points are classified perfectly, and one gets a wide margin. Nevertheless, one pesky data point is left that cannot be placed on the margin's right side: it may be noise. One wants to pretend that the point can be "moved" to the right side through the hyperplane. To do so, one will have to pay a little bit, but if one is not shifting many points around, doing this should be a good idea. The amount that one shifts the point in this way is denoted $\xi(x_i)$. One can construct an objective function like the following by adding one slack parameter for each training example and penalizing oneself for having to use slack:

$$\min_{w,b,\xi} \frac{1}{2} w \cdot w + \sum_{i=1}^{n} \xi_i, \text{ such that } y_i\left(w \cdot x_i + b\right) \geq 1 - \xi_i, \,\& \,\xi_i \geq 0 \quad (6.18)$$

This slack feature aims to ensure that all points are properly categorized (the first constraint). Nevertheless, if one cannot properly categorize point i, one can set the slack ξ_i to anything greater than zero to shift it in the right direction. However, for all nonzero slacks, one must pay equal to the amount of slack in the objective function. Overfitting versus underfitting is regulated by the hyperparameter $C > 0$. Notice that a trade-off justification achieves optimization; the user-defined parameter, C, governs the effect of each of the two contributions to the task of minimization. The resulting margin (the distance between the two hyperplanes) will be small if C is large so that a smaller number of margin errors are committed. The opposite is true if C is small. The corresponding Lagrange function is:

$$\mathcal{L}\left(w,b,\mu,\alpha,\xi\right) = \frac{1}{2} w \cdot w + C\sum_{i=1}^{n} \xi_i - \sum_{i=1}^{n} \mu_i\left[y_i\left(w \cdot x_i + b\right) - 1 + \xi_i\right] - \sum_{i=1}^{n} \alpha_i \xi_i$$

$$(6.19)$$

where $\alpha_i, \mu_i \geq 0$ are Lagrange multipliers. To find minima, we differentiate the Lagrange function in Equation 6.19.

6.7 THE KERNEL FUNCTION

In ML, kernels are used to convert the data for achieving more straightforward classification. Data in the lower dimensions are difficult to classify. As we increase the number of classes and data points, the boundary line between two sets of data classes becomes more complicated. Therefore, we need to translate these data into higher dimensions to distinguish different groups by drawing simple hyperplanes. A matrix represents the input data vector x_i in the dual formulation of the support vector machine problem. In other words, in the dual formulation, only the input vectors' internal products play a role – there is no explicit use of x_i, or any other function of x_i, other than the internal products. This observation indicates the use of what is known as the "kernel function" to substitute nonlinear functions for inner products. Constructing nonlinear versions of linear algorithms by replacing inner products with nonlinear kernel functions is a common concept of kernel methods. Under some conditions, this approach can be interpreted as mapping the original measurement vectors (so-called "input space") to some higher-dimensional (possibly infinitely high) space generally referred to as "feature space."

Now let us look at this theory and explain why the kernel trick works, including what it means to function in an infinite dimension? Let us begin with the most common instance and then extend to the general case – an example of the original data being in two dimensions as seen in Figure 6.11. In Figure 6.11a, we can see that data groups are inseparable within their space. If we find a way to map the data from two-dimensional space to three-dimensional space, we will be able to find a decision field that distinguishes explicitly between various groups that can be segregated in a transformed space (Figure 6.11b) given by:

$$\Phi(x) \rightarrow x_1^2, x_2^2, \sqrt{2}x_1x_2 \tag{6.20}$$

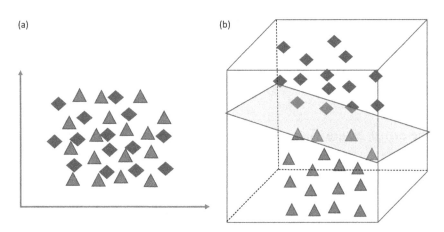

FIGURE 6.11 Example of labeled data that are (a) inseparable in two dimensions but are separable (b) in three dimensions.

where Φ is a 2-D to 3-D transform function applied to x. We will now have a decision boundary in a 3-D space with the Φ in Equation 6.20 that will look like:

$$\beta_0 + \beta_1 x_1^2 + \beta_2 x_2^2 + \beta_3 \sqrt{2} x_1 x_2 = 0 \tag{6.21}$$

If we were doing a logistic regression, our model would be like Equation 6.21. We cannot just arbitrarily project it as we attempt to transform data into higher dimensions. We do not want the mapping to be directly computed either. Therefore, to develop the boundary, we use only specific algorithms that use vectors' dot products in the higher dimension. We do not necessarily have to project the data into higher dimensions to compute vectors' dot products in the higher dimension. Using the lower dimension vectors, we will use a kernel function to compute the dot product directly. Those candidate linear algorithms are converted into a nonlinear algorithm when correctly implemented onto the lower dimension. Placing a curved border would become equal to putting a linear boundary in the higher dimension.

One crucial point to bear in mind, though, is that there are chances that we can overfit the model when we map data to a higher dimension. Choosing the correct kernel function (including the correct parameters) is also very critical. Different kernels exist. The polynomial kernel and the radial basis function (RBF) kernel are the most common ones. A kernel's preference depends on the topic at hand, so it depends on what we are trying to model. For example, a polynomial kernel enables one to model conjunctions of features up to the polynomial degree. In contrast to the linear kernel, which can only select lines (or hyperplanes), radial basis functions can select circles (or hyperspheres).

CHAPTER 7

Prediction, Optimization, and New Knowledge Development

A MIXTURE OF MACHINE LEARNING (ML) and optimization is used to make many real-world analytics applications in research and science. Different methods have been developed, ranging from cost function optimization, which tries to minimize the deviation between what is observed from the data and what is predicted by the model, to probabilistic models that try to model the data's statistical characteristics. The optimization model is usually used to make decisions. In contrast, a ML instrument is used to generate a simulation model that forecasts the optimization model's core unknown parameters to predict the outcome. Consider a vehicle routing problem, for instance, that can be solved many times a day. Based on actual traffic, temperature, holidays, and time, a previously trained prediction model makes forecasts for the travel time on all edges of a road network. Then, using the expected travel times as data, an optimization solver offers a near-optimal route. Some aspects of both estimation and optimization are involved in most solution systems for real-world analytics problems.

Currently, developing solutions is a synergistic mixture of ab initio calculations, simulation, ML, and optimization techniques for the classification, resolution, and prediction of scientific evidence and associated phenomena. These tasks include rapid literature searches and reviews through historical data analysis techniques, design of experiments for

DOI: 10.1201/9781003206743-7

physical and quantum physical property analysis, and transfer function development for each parameter of interest. Estimation and prediction can be achieved through neural networks leading to digital twin development. Optimization is achieved through simulations leading to new products, applications, information technology advances, and new knowledge development. The most effective and cost-competitive perpetual cycle for knowledge development is shown in Figure 7.1.

The generalization of scalability of broader problems is now the core theory for transforming tasks on several fronts. Systemic and cost-effective methods, including those focused on deep learning, have gained traction from ML approaches. The latter class of ML algorithms can combine raw input into intermediate function layers, allowing designs of bench-to-bytes with the ability to convert several domains. Simulation efficiently and successfully resolves real-world concerns. It offers a basic form of research that is readily checked, transmitted, and understood. Digital twins are developed for simulation modeling that offers useful solutions across sectors and fields by providing simple insights into dynamic processes.

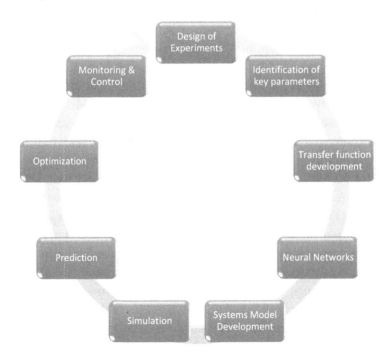

FIGURE 7.1 Typical cycle for prediction, optimization, and new knowledge development involving digital twins and cyber physical systems.

7.1 DIGITAL TWINS

The emergence of autonomy is imminent, but it is still very recent. It is increasingly growing in the transportation sectors. There are vehicles, planes, and boats capable of driving themselves, but society does not feel secure enough to turn to unmanned systems entirely in these cases. The growing volume of data available is used to analyze better and optimize the various aspects of an unmanned, autonomous system. When the planet becomes more connected, and technologies can capture and store more significant and larger volumes of data, an opportunity opens for too complex machine simulations to be made. These simulations are not solely based on first-principle physics models but rather take a vast amount of data from the devices' sensors. Furthermore, integrate them to explain them even more clearly and develop a closer to the system's real picture with the current models. These simulations are intended to be continuously linked to the physical system and simulate any part of the physical system in real time.

A digital twin is described as a digital replica of physical assets, processes, and systems that can be used for different purposes. A digital twin can be thought of as a bridge between the real and the digital universe. The digital twin conceptual paradigm builds on and digitizes the physical system and uses digital language to construct a virtual product of the same substance and structure as the physical entity's appearance; it incorporates virtual reality and creates the connection between virtual space and real space. It visually represents the notion of combining the real with the images so that data and information can be shared between each other (Figure 7.2). It is an immersive, multiscale, probabilistic multiphysics simulation of an as-built system that is enabled by a digital thread. This simulation mirrors and forecasts its lifetime corresponding to physical twin behaviors and output using the best available models, sensor information, and input data. It makes it possible to monitor the real with the imaginary. Furthermore, it is possible to set up the corresponding digital twin model in the virtual environment, in addition to goods, for warehouses, workshops, production lines, manufacturing tools (workplace, machinery, staff, and materials).

In theory, the physical twin integrates the design of holistic simulation models with the understanding of the efficiency of actual machines and processes. A data monitoring infrastructure must be incorporated to comply with this second aspect so that needed information is correctly acquired, handled, and analyzed. The method used in twin control consists of installing local tracking hardware that acquires the machine's

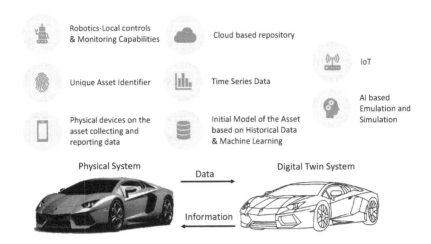

FIGURE 7.2 The concept of digital twin illustrated with its key components and processes. AI, artificial intelligence; IoT, Internet of things.

internal variables, gathers additional sensor information, and uploads all data to a cloud server. A fleet-level data review is carried out in the digital twin by combining all the various machines' information.

Cyberphysical systems can realize the relationship between the virtual model of data and the physical entity by creating a closed-loop channel for data interaction between information space and physical space. The advent of the digital twin provides cyberphysical systems with a simple meaning, process, and execution method. It consists of the physical entities in the physical space that are reconstructed in the information space. This reconstruction employs real-time data acquisition, data integration and monitoring, dynamic tracking of physical entity working status, and progress assessment results (such as acquisition and traceability information). Therefore, from this angle, the primary technology of cyberphysical systems is the digital twin.

Cyberphysical systems have become the core technology for Industry 4.0, a term repeatedly used for the next industrial revolution manifesting through current times. A cyberphysical system is a dynamic multidimensional system integrated with software, communication, power, network, and physical environment. The real-time perception, dynamic control, and knowledge service of large-scale engineering systems can be realized based on the big data network and mass computing and the organic integration and deep cooperation of computing, communications, and control technologies.

In the future, there will be a digital twin construct in virtual space that is the same as an object in physical space. For example, in virtual space, the physical factory has a corresponding digital twin factory model, and in virtual space, the physical workshop has a corresponding digital twin workshop model. In virtual space, physical production lines have the corresponding digital twin model of the production line. The digital twin is the cornerstone of a smart production system. The digital twin's most crucial enlightening value is recognizing the physical system's input to the digital cyberspace model.

7.1.1 Advantages of Digital Twins

As digital twin is not a particular technology, all the advantages this term offers are difficult to explain. Many digital twins, however, have capabilities that have the following usual advantages:

Visibility: The digital twin has visibility in processes and larger integrated structures such as a production unit, a warehouse, a financial system, or an airport.

Predictive: The digital twin model can be implemented using different modeling approaches (physics-based and mathematics-based) to forecast the potential state of the devices and optimize characteristic output parameters, from energy usage to error rates, repair, and cleaning cycles.

Experimental prototypes (what-if analysis): It is easy to communicate with the model by correctly designed interfaces and ask what-if questions and design tests to the model to replicate different impossible or costly situations to construct in real life.

Documentation and information process to understand and describe behavior: A digital twin may be used to understand and explain an individual computer's behavior or a compilation of data from many related devices as a communication and documentation mechanism.

Link various structures such as business applications in the backend: If appropriately built, the digital twin paradigm can connect with business applications to obtain business results in supply chain activities, including manufacturing, sourcing, warehousing, transport, logistics, and field service.

7.1.2 Development of Digital Twins

As envisaged by Industry 4.0, the Factory of the Future heavily relies on digital twins. Digital twins help manage manufacturing operations and product life cycles and many use cases that can be defined as forms of tracking, optimization, and control. Current technology has two limitations: the absence of applicable techniques for modeling human actions and their interaction with machines. The closed-system reality of existing digital twins restricts their ability to provide holistic mechanisms for optimizing processes.

To evaluate solutions to a problem at hand, many approaches require historical data. These approaches include k-nearest neighbor, support vector machines, artificial neural networks, and other techniques in artificial intelligence. For diverse industrial problems, these approaches have achieved outstanding results. Nevertheless, they need extensive training data that incorporate diverse types of problems that add to the challenge.

To start, mathematical models can represent all dynamic systems. These models regulate differential equations in continuous time. Furthermore, since many systems are only sampled at a given rate, it is often useful in discrete time to analyze systems as differential equations. For linear systems, transfer functions define the input–output relationship between certain variables in the Laplace domain. The relationship between inputs and outputs is defined as a polynomial relation by a linear autoregressive model. State-space models explain the state variables' evolution over time. The dynamics between inputs and outputs are represented with nonlinear polynomials using nonlinear autoregressive models for nonlinear systems. Hammerstein and Weiner models (commonly used to construct adaptive control systems for nonlinear stochastic dynamic objects of various kinds) integrate inputs and outputs of linear transfer functions with nonlinear functions.

These methods are well adapted for the digital twin paradigm since historical data can be continuously obtained while in service and used to refine artificial intelligence models. The artificial neural network is one particular case of knowledge-based approaches. Typically, these networks are used for image processing but can be extended to a variety of issues. Each network node has a certain number of inputs and a single output (which can be sent on to several other nodes). To evaluate the output, these inputs are weighted and summed up. In specific networks, the nodes include an activation function that, depending on the weighted sum's value, converts the output to zero or one (Figure 7.3).

Usually, nodes are divided into layers, and the outputs from one layer typically become the inputs to the next layer. The output layer that holds the information useful to the analysts is the final layer of every network. By changing the weighting of the inputs at each node, networks are trained.

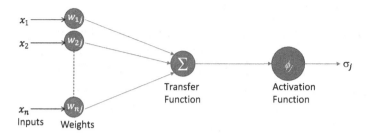

FIGURE 7.3   Neural network node. Transfer function is often a summing function.

Several training algorithms modify these weightings based on training data supplied to the network. A set of inputs and their valid, equivalent outputs is training knowledge. The training algorithms change each node's weights such that each set of inputs better obtains the right set of outputs.

Sometimes, one may derive the structure of the model from first principles. Then, one needs to derive the constants within the model. This method is called the estimation of parameters. To evaluate these constants, valid input and output data from experiments are required. The Device Identification Toolbox usually includes guidance to ensure that the data are of sufficient quality to capture all the system's driving dynamics. These include data with a high signal-to-noise ratio. These data agitate the system in multiple ways (i.e. more than a single-phase input), data that last long enough to stabilize the system that is sampled at acceptable rates. For a standard set of input data, parameter estimation algorithms tune the parameters to minimize the discrepancy between the model output and the measured data output. Usually, this method involves minimizing a particular function that is often a set of weighted totals of square errors. In the tuning of the model parameters, this function determines the importance of some outputs over others.

Model testing is the final stage of device recognition. Techniques for assessing the model's output are included in the Device Recognition Toolbox. Second, the model response is contrasted with the measured response. This comparison creates a percentage fit, which shows how well the model fits the data. A great fit (which may be overfitted and the following noise) is shown by 100%, while 0% reveals an inferior fit. Secondly, the residual analysis explores how the model is linked to past inputs. Thirdly, for each parameter value, model uncertainty analysis defines a confidence level. For the model, this creates a measure of precision that can be interpreted in the time or frequency domain. After testing the model using these techniques, it is essential to decide whether the model is accurate enough for its purposes or not. If not, the model's configuration, the input/output of the data, or the criteria function must be modified. The algorithm of parameter estimation must be rerun before convergence to an appropriate model.

To shorten production cycles, accelerate the launch of new products, and minimize process inefficiencies, digital twins thus continuously collect data and analyze process information. Besides, digital twins are linked to optimization, with a digital twin being a simulation with the ability to monitor goods and production lines in real time and optimize them. As the generally accepted definition of digital twins is not yet available, it is not surprising that there is a lack of digital twins' construction requirements. In the emerging literature, however, there are many approaches available to modeling and constructing digital twins that seek to assist the system engineer in the digital twin construction process.

These primary enabling approaches are grouped into five categories in the literature: physics-based modeling, data-driven modeling, cybernetics of big data, networks, and platforms, and human-machine interface. Physics-based simulation techniques explore how hypotheses formed by studying physical phenomena can be converted into mathematical equations and how to solve them. 3D modeling and simulation techniques and numerical solvers that are already built into numerical simulators are examples of such technologies. Data-driven modeling aims to detect trends in large data sets, assuming that information on physical phenomena is already encoded in the data.

Enabling technologies also include techniques for generating data, algorithms for ML, and artificial intelligence approaches. These techniques make it possible to minimize the gap between the actual operating situation facing the real device and the interpreted situation generated by the digital twin through its physical counterpart's sensor data. Big data cybernetics incorporates the theory of control with big data and attempts to direct the machine to a given state. The system's behavior is continuously tracked from control theory, and evolution is compared with the reference state. Big data methods are used to enhance awareness of the observations, which can be used for the digital twin and its physical equivalent to boost controller performance. IoT, cloud computing, and 5G are the enabling technologies for digital twins.

Based on industrial organizations' experience, the envisioned future suggests that the development, control, and monitoring of intelligent and connected goods would transition from human labor-centered production to fully automated production. In this regard, Industry 4.0 transformation includes strategic preparation of the workforce, creating the right organizational framework to create alliances, and engaging in and sharing technical standardization, which are essential factors for driving technological change. Real-time supply chain optimization, human–robot collaboration, smart energy use, automated performance control, and predictive maintenance would be significant implementation areas in manufacturing supported by digital twins. Furthermore, by adapting nanotechnology and robotics to the implementation of Industry 4.0, supporting technologies can be more successful. Besides, self-organized, self-motivated, and self-learning systems will be expected soon with more advanced artificial intelligence algorithms and autocreation of business processes.

7.2 MONTE CARLO SIMULATIONS

When the data are obtained, it can construct computer simulation models for organizational effects, predict conditions and outcomes, and assess behavior. The model may vary from a basic three-variable equation (e.g. $A+B=C$) to a highly complex and too complicated set of interconnected formulas and spreadsheets. Creating the right model requires time, persistence, strategy, and practice. Once built and validated, these digital twin models will recommend behavior based on engineering simulations, physics, chemistry, statistics, ML, artificial intelligence, and business goals. These models can be shown by 3D representations and improved simulation of reality to help people comprehend the effects.

Simulation opens the door for the practitioner to solve complex but realistic problems quickly. The probability theory enables the simulation of complex phenomena whose states cannot be reliably deduced from precise measurements. Classical approaches cannot accurately analyze such phenomena. Stochastic modeling consists of selecting the data's probability distributions and measuring the distributions of probability of the phenomenon's significant characteristics under consideration. Monte Carlo is such an approach that is usually associated with the modeling and simulation phase of a randomized system: multiple random scenarios are generated, and related statistics are collected to determine, e.g. the performance of a decision policy or the value of an asset. Monte Carlo techniques are highly versatile and useful methods for excessively complex issues impervious to a more mathematically elegant treatment. This simulation type is commonly used in many different fields, including engineering, physics, research and development, industry, and finance. Monte Carlo simulations generate artificial futures by creating thousands, even hundreds of thousands, of results and analyzing their prevailing features.

A Monte Carlo simulation calculates several model scenarios by repeatedly sampling values for the unknown variables from a user-defined probability distribution and using those values for the model. Since all these scenarios generate the results in a model, a forecast can be given for each scenario. Projections are events, which one describes as important model outputs (usually with formulas or functions). Think of the Monte Carlo simulation approach as the repetitive pickup of apples from a large bag. The size and type of the bag depend on the assumption of the input distribution. A normal distribution with mean of 100 and standard deviation of 10, for example, versus a uniform distribution or a triangular distribution. In later cases certain bags are more profound or symmetrical than others, allowing some apples (gala, red delicious, or granny smith) to be extracted more often than others. The number of apples drawn repeatedly depends on the number of simulated trials.

Imagine the large model as a huge bag, where several baby bags live, for a large model with many associated assumptions. Each baby bag has its collection of apples of different kinds. These baby bags are often connected (if there is a connection between the variables), causing the apples to jump in tandem. In contrast, the apples shift independently of each other in other uncorrelated instances. The apples selected each time within the model (the large bag) from these interactions are tabulated and registered, providing the simulation's forecast performance result.

Even though a random variable's value cannot be anticipated, the distribution can be well defined. A random variable's distribution gives the likelihood of a given value. A discrete random variable's probability distribution lists the probabilities associated with each of its possible values. It is often called the probability distribution function or the probability mass function, which have been described and discussed in detail in Chapter 4. The mean, or the expected value function, and the variance were also described in detail in Chapter 4. Given two independent variables, x and y, the covariance and correlation of x and y are defined as:

$$Cov[x,y] = E[xy] - E[x]E[y] \qquad (7.1)$$

$$Corr[x,y] = \frac{Cov[x,y]}{\sqrt{Var(x)\,Var(y)}} \qquad (7.2)$$

Note that if x and y are uncorrelated, their COV and CORR will be zero. The multiplicative congruential process is one of the common algorithms used in Monte Carlo simulations. Using the following equation, this method generates successive pseudo-random numbers given modulus "m" and multiplier "a" with a starting point x_i :

$$x_i = ax_{i-1} \bmod (m) \tag{7.3}$$

Let $f(x)$ be a continuous arbitrary function, $y=f(x)$. Given by an arbitrary interval (a, b), the expected value $E(y) = E[f(x)] = \int_a^b f(x)p(x)dx$, where $p(x)$ is the probability of x. Without computing the integral, Monte Carlo simulations assist us to calculate $E(y)$. It is possible to obtain a simple integral estimate by generating n samples $x_i{\sim}q(x)$ for $i = 1, 2, ..., n$ and by computing the estimate:

$$E(y) = \frac{1}{n}\sum_{i=1}^{n} f(x_i) \tag{7.4}$$

This method's precision derives from the law of large numbers and the central limit theorem. The law of large numbers states that, as the number of samples goes to infinity, the average of a random variable sequence of a known distribution converges to its expected value. Furthermore, the central limit theorem states that approximately the sum of many independent variables is uniformly distributed. The Monte Carlo simulation system error is therefore given as:

$$\text{Error} = \left|E[y] - E[f(x)]\right| = \sqrt{\frac{Var[f(x)]}{n}} \tag{7.5}$$

Increasing the sample size increases the accuracy of the Monte Carlo simulation, but convergence is very slow. Decreasing the variance is a safer approach to improving precision.

It is stressed here that the theory of probability makes it possible to model complex phenomena whose states cannot be reliably deduced from precise measurements. Classical approaches fail to study such phenomena. Stochastic modeling consists of selecting the data's probability distributions and measuring the distributions of probability of the phenomenon's significant characteristics under consideration. Statistical information on the system under consideration is required: the mean values of speed, location, resources, price; the probabilities of critical incidents, such as long cracks, significant financial losses, large numbers of people infected with an epidemic; the density of particle positions and speeds, the distribution of probability for neuronal firings.

The distribution of probability, expectation, or probability of critical events needs to be computed in both cases. By solving deterministic equations, the preceding quantities can often be obtained. For instance, contingent claims' prices can often be expressed in terms of parabolic partial differential equation solutions. When the state space dimension is sufficiently small, these partial differential equations' numerical resolution by deterministic methods is possible (and recommended). The numerical complexity of deterministic methods usually explodes when the state space dimension increases. Stochastic numerical methods are used to obtain quantitative data on probabilistic models and to solve deterministic equations, including equations that directly describe deterministic phenomena. The Monte Carlo simulation is a parametric simulation in which specific distribution parameters are needed before a simulation can begin. A nonparametric simulation is a tool for conveying a narrative using raw historical data without the use of distributional parameters.

7.3 RESPONSE SURFACE METHODOLOGY

The response surface approach is one of the methods for the design of experiments used to estimate an unknown function for which only a few values are computed. The response surface methodology derives from science disciplines. Physical experiments are carried out to research the unknown relationship between a series of variables and the system's output or response. Only a few values of the experiments are acquired. Using a mathematical model called the response surface, these relations are then modeled. Response surface methodology can be defined as a collection of mathematical and statistical processes useful for modeling and analyzing problems where many variables function in the responses or outputs. Figure 7.4 gives an example of a response surface. The exact relationship between the response and the variables or variables is generally not explicitly understood, and an approximate empirical correlation or function is used to match a relationship. Two types of functions are typically used to estimate this relationship: a first-order empirical model or a second-order model. Good experimental designs may derive the coefficients in these models.

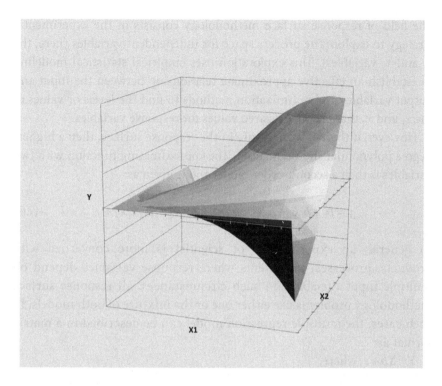

FIGURE 7.4 Arbitrary response surface for the output variable y, that is dependent on two input (independent) variables x_1 and x_2.

Once a reasonably acceptable model based on the available data is built using the theory and methodology given in Chapter 4, response surface methodology can provide predictions for the optimization. The response surface methodology aims to optimize a response (output variable) influenced by several independent variables (input variables). Let us consider a simple case where the response variable "y" is dependent only on two variables x_1 and x_2; then we obtain $y=f(x_1, x_2)+\varepsilon=b_0+b_1x_1+b_2x_2+\varepsilon$, where the error term ε represents any measurement error on the response, as well as other type of variations not counted in f. Notice that there is a corresponding output value y for each value of x_1 and x_2 and that we can interpret these response variable values as a surface above the plane of x_1-x_2, as shown in Figure 7.4. It is this graphical viewpoint of the problem that contributed to the term response surface methodology.

It would be straightforward to refine this method if we could conveniently create the graphical displays in Figure 7.4. Unfortunately, the correct solution variable in Figure 7.4 is unclear in most realistic circumstances.

The field of response surface methodology consists of the experimental strategy to explore the process space for independent variables (here, the x_1 and x_2 variables). This exploration uses empirical statistical modeling to establish an effective approximate relationship between the input and output variables and optimization methods to find the levels or values of the x_1 and x_2 that produce desired values for response variables.

However, if there is a curvature in the response surface, then a higher-degree polynomial should be used. The approximating function with two variables is then a second-order model and is given as:

$$y = b_0 + b_1 x_1 + b_2 x_2 + b_{11} x_1^2 + b_{22} x_2^2 + b_{12} x_1 x_2 + \varepsilon \qquad (7.6)$$

In general, an experimenter or scientist is more concerned with products, processes, or systems where response variables depend on multiple input variables. In such circumstances, all response surface methodology problems use either one or the mixture of both models. In such cases, the multiple regression model can be described in a matrix format as:

$Y = Xb + \varepsilon$, where,

$$Y = \begin{vmatrix} y_1 \\ y_2 \\ \vdots \\ y_n \end{vmatrix}, X = \begin{vmatrix} 1 & x_{11} & x_{12} & \cdots & x_{1p} \\ 1 & x_{21} & x_{22} & \cdots & x_{2p} \\ \vdots & \vdots & \vdots & \vdots & \vdots \\ 1 & x_{n1} & x_{n2} & \cdots & x_{np} \end{vmatrix}, b = \begin{vmatrix} b_0 \\ b_1 \\ \vdots \\ b_p \end{vmatrix} \text{ and } \varepsilon = \begin{vmatrix} \varepsilon_1 \\ \varepsilon_2 \\ \vdots \\ \varepsilon_n \end{vmatrix}$$

$$(7.7)$$

Simulating Equation (7.7) generates multidimensional response surfaces (a two-dimensional response surface is shown in Figure 7.4) that can be used to optimize the system based on multiple response variables.

The b's are a set of undefined parameters. We must gather data on the system we are studying to approximate the values of these parameters. Regression analysis is usually used to construct a statistical model from these data to estimate the b's. Since polynomial models are usually linear functions of the unknown b's, we refer to this as linear regression. We can also see that carefully preparing the data collection process of a response surface methodology study is very necessary. In this regard, particular types of experimental designs are useful, called response surface designs.

7.3.1 The Sequential Application of Response Surface Methodology

The most useful applications of response surface methodology are sequential. Some ideas are first created about the variables or factors that are likely to be important in analyzing the response surface. This variable selection generally leads to experiments designed to analyze these variables to remove the unimportant ones. This type of experiment is generally called a screening experimental design. There is always a very long list of variables at the beginning of a response surface analysis, which may be useful in describing the response. Factor screening aims to reduce to a relatively few but essential list of candidate variables such that subsequent experiments are more effective and require fewer runs or tests. This stage aims to identify the most important independent variables impacting the output response variable.

Phase 1 of the response surface analysis starts once the relevant independent variables are identified. At this point, the experimenter aims to decide if the independent variables' current levels or settings result in a response value similar to the optimum or if the process operates in some other region that is (possibly) remote from the optimum. Suppose the current settings or levels of the independent variables are not compatible with the desired optimum. In that case, the experimenter will decide a series of changes to the process variables that will shift the process toward the optimum. This step of the response surface methodology makes extensive use of the first-order model and an optimization technique called the steepest ascent or descent process depending on the optimum being maxima or minima, respectively.

When the process is near the optimum, phase 2 of a response surface analysis starts. The experimenter typically needs a model at this stage that will estimate the correct response function reliably within a relatively small area (design space) around the optimum. Since the proper response surface typically shows curvature near the optimum, a second-order model (or some higher-order polynomial, occasionally) is used. This model is evaluated to determine the method's optimal conditions until an adequate estimated model has been obtained. This sequential experimental method is typically carried out within an independent variable space area called the region of interest or the operability region.

The sequential design technique for the response surface methodology enables the experimenter to learn as the inquiry continues about the mechanisms or systems under analysis. This technique ensures that the experimenter can learn the answers to questions such as (1) the position of the optimum area, (2) the type of approximate function required, (3) the correct range of experimental designs, (4) how much replication is needed, and (5) whether changes to the responses or any of the process variables are required.

7.3.2 Robust Design

Methods of multidisciplinary optimization have become particularly important for improving design efficiency and cost savings in engineering problems. The virtual prototyping method is interdisciplinary. Such a multidisciplinary approach involves the simultaneous operation of different solvers and managing different types of constraints and goals. There is a need to link arbitrary engineering and complex nonlinear analyses. The resulting optimization problems may become very noisy, very sensitive to design changes, or poorly conditioned for mathematical function analysis (e.g. nondifferentiable, nonconvex, nonsmooth).

For the past few decades, there have been many obstacles to virtual prototyping. "Built-in-quality" and "built-in-reliability" are more and more the focus of engineering. Products must be produced in the shortest possible time and must be safe, durable, and resilient in lieu of that. Simultaneously, numerical models are getting more and more comprehensive, and numerical procedures are becoming more complicated. For numerical analysis, considerably more reliable data are needed. These data are usually random parameters. The optimization method entails uncertainties or stochastic dispersion of design variables, objective characteristics, and constraints that must be carefully handled.

In comparison, optimized architectures or designs contribute to high susceptibility to imperfection and appear to sacrifice robustness. The deterministic optimum design is often moved to the design space boundary using a multidisciplinary optimization process. There is no room for tolerances or uncertainties in the design properties. It would also be more and more critical to determine design robustness, durability, and safety. Therefore, it is essential to combine optimization and stochastic methods of design analysis.

In essence, robust parameter design is a concept that emphasizes the correct choice of levels of controllable factors in a system. A process or a product may be the system under consideration. The theory of selection of levels focuses primarily on uncertainty around a preselected performance response target. These controllable variables are called factors of influence. The bulk of deviation around the target is caused by the presence of the second group of variables called noise variables. In the design of the product or the device's regular function and processes, noise factors are uncontrollable. It is essentially this absence of control that transmits heterogeneity to the response of the system. Consequently, the term robust parameter design involves engineering the device (not in the context of experimental design) to achieve robustness (insensitivity) to unavoidable changes in the noise variables (selecting the levels of the controllable variables).

In the experimental design, simulation, and optimization exercise, a large portion of the methodology focuses on reducing process variability by involving variables that affect process variability. The noise variables are also functions of ambient factors, such as temperature, raw material properties, and items' age. These may include the way the end user treats or uses the device or method of some applications. Noise variables can be, and sometimes are, managed at the level of research or development, but they cannot, of course, be controlled at the level of manufacturing or product use.

For robust design, two design rules offer a structure. The basic postulates are stated as follows: (1) The independence postulate states that the functional requirements (FRs) must maintain independence and (2) the information postulate is to minimize the design's information content. The postulate of independence states that an acceptable choice of specification parameters must preserve the design parameters' (DPs') independence. The information postulate provides a metric for choosing the right design when the independent postulate is satisfied by several designs. By minimizing data information, the best value can be achieved for chosen DPs. Information content I is defined as:

$$I = \log_2 \frac{1}{p} \qquad (7.8)$$

in terms of the probability p of meeting a given FR. The content of information is specified in terms of probability of performance, determined by defining the range of design and systems range. The data are specified as:

$$I = \frac{A_{cr}}{A_{sr}} \qquad (7.9)$$

where the common range is A_{cr}, and the system range is A_{sr}. To minimize the information content, (1) the system variation must be minimal, and (2) the bias must be eliminated to render the system range within the

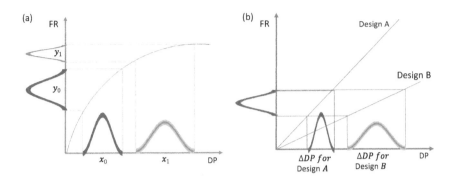

FIGURE 7.5 Dependence of functional requirements (FRs) on design parameter. (a) The concept of the gradient of the transfer function and its relationship to the variation transmission from the inputs to the outputs (b) choosing a design with lower gradient provides the robustness of the product or the process. DP, design parameter.

design range. That is, the minimization of the content of information is analogous to the robust conception of design. Satisfying the independence postulate allows the development of a robust design by minimizing each FR's information. Therefore, it follows that there is one FR one DP:

$$FR = A \cdot DP \qquad (7.10)$$

This dependence is simplified by a specific design parameter (DP) in Figure 7.5a. The gradient of the so-called transfer function shown illustrates the product's susceptibility to the input parameter variance, that is, the robustness. Suppose there are two candidates' DPs in Equation 7.10 and that the FR-DP relationships are seen in Figure 7.5b. A design's gradient is considered the design's stiffness. The defined FR tolerance can be accomplished more effectively by reducing the stiffness of the design (design B). Allowing DP to be larger contributes to a robust design.

7.3.3 Simulation Tools for Response Surface Methodology

In a physical process, such as chemical vapor deposition in semiconductor manufacturing, wave soldering, or laser machining, we typically think of applying response surface methodology to understand multidimensional relationships. Response surface methodology can also be effectively extended to physical structures modeled by computers, like digital twins. These digital twin models are becoming more and more popular and can be used as proxies for complicated procedures that are difficult or costly to manipulate. In specific applications, due to cost, protection, or legislation, there are limitations on what conditions can be tested for a physical experiment. In other examples, having a programming model (or code) enables the discovery of alternatives, concept production, and new technologies to be even more comfortable, increasing comparative advantage and business speed. Computer models may also have many inputs, leading to multiple responses that can either be scalar (a single value) or functional (a collection of values connected over time or space). Therefore, the experimental design has an important role to play in selecting a preferred set of combinations of inputs to explore.

The function of response surface methodology is different in such computer modeling applications. The data obtained from runs of the computer model can generate a model of the device modeled by computer simulation, called a metamodel emulator. As it can take multiple program runtime to extract one observation from the code for specific computer models, knowledge of the system's characteristics can be gleaned from exploring the approximate model and optimizing the metamodel. The premise is that if the computer simulation model is a valid representation of the real system, then the optimization of response surface methodology would result in the optimal conditions for the entire system being adequately determined.

Two types of simulation models, stochastic and deterministic, are commonly available. The response outputs are random variables in a stochastic simulation process. Examples include machine simulations for factory planning and scheduling models used by civil engineers in the semiconductor industry and traffic flow simulators. Furthermore, Monte Carlo simulations that sample complex mathematical processes do not have straightforward empirical solutions from probability distributions to study them.

The output responses are not random variables in deterministic simulation models. They are wholly deterministic quantities whose values are defined by the (often too complex) mathematical models on which the computer model is based. Deterministic simulation models are frequently used as computer-based modeling methods for engineers and scientists. Circuit simulators used to construct electrical circuits and semiconductor devices, mechanical and structural design finite element analysis models, and theoretical models for physical processes such as fluid dynamics are common examples. Sometimes, these are very complicated simulations that take substantial computing resources and time to run.

It helps compare data collected from physical experiments and computer models when deciding what designs and analyses to use when analyzing computer models. Data from all types of experiments can be costly: The expense comes from setting up and conducting the experiment with physical materials and calculating the response values. The creation of codes is labor- and time-intensive for computer experiments and needs subject matter skills. We presume that the computer model is already usable, and we want to investigate the fundamental relationship of interest by using it. It is always costly to collect the data itself. Due to the code's sophistication, even with a moderate number of input variations, it may take significant computing power and runtime to achieve results.

Therefore, achieving a reasonable approximation of the relationship effectively from small to moderate quantities of data is essential for all kinds of experiments. The spectrum over which experimentation can take place is a crucial distinction between physical and machine experiments. For physical experiments, the purpose is always to limit the area of interest to a local region where the relationship between inputs and outputs well defines a low-order polynomial. However, an experiment should be pursued for computer simulation experiments to describe the relationship over even broader input space regions. The physical constraints often enforced by where admissible values can be obtained in the operability region are not present. The limits where insights from programming codes can be obtained are always determined by evolving science or engineering processes that have been designed into the code.

Physical experiment evidence reflects findings from the real phase and is usually considered neutral, based on using a sound measuring system to calculate the response output. However, we usually do not intend to find similar values for repeat runs of the experiment due to defective measuring instruments, minor variations in how the input settings might be set, and the procedure's inherent variability. In this way, to catch and approximate these variations, our mathematical models for physical processing are characterized with an error word.

On the other hand, evidence from computer simulations, based on the best available science and engineering of the processes driving the operation, results from a human-made characterization of the interaction. These codes can vary from very precise to only coarse representations of the critical structures, based on the code's maturity and the extent of underlying information. As a result, the user should be mindful that any project's outcomes to explore a programming model are just as strong as the simple information available for the code to be created. Therefore, there is a risk of bias or systematic inconsistencies in the code and the true mechanism it is supposed to portray.

Therefore, if the model is deterministic, similar values would result in repeated runs of the same inputs, rendering this function undesirable for this form of machine experiment. Any approximate metamodel or simulator for the code to interpolate between observed points is useful. Since we agree that the code's findings reflect the best interpretation of the underlying process possible, we want to use these data explicitly. The simulator's purpose is to enable other positions in the input space that has not been measured explicitly to be estimated. If the computer model is stochastic, it could be necessary to acquire replicates to understand the natural variability.

The fundamental interaction being characterized is another key factor to remember when comparing physical experiments and computer models. Response surface methodology is based on the premise that certain physical processes are smooth and continuous and can be well approximated by low-order polynomials, at least in the region of interest. For certain complex programming codes, this inference may not be suitable. Polynomials of a higher order than the normal quadratic reaction surface models, for example, are often used in such complex cases. Since various options are often appropriate to explain the underlying relationship, some advanced techniques may be available for planned experiments based on different models.

The variations between the actual model based on the physical experimental data and computer simulation models are shown in Figure 7.6. Consider a very simple scenario where we are calculating an output from a single input factor. Because the underlying assumptions about the data vary, different model choices and related uncertainty can be obtained. For the physical experiment data in Figure 7.6a, a linear regression line with predicted intercept and slope is the assumed model. The magnitude of each deviation is used to approximate the normal variance in the output, as the measurements do not lay exactly on the best-fitting line. If, on the other hand, as seen in Figure 7.6b, the data come from a deterministic simulation experiment, then an interpolator model is chosen. The best approximate value for input is matched with a data point acquired from the machine code. As we have an exact value at that position for the computer code results, the simulator's variance is zero, and the uncertainty precisely around the function shrinks to zero. Note that as we step away from any observation, the uncertainty about the predicted curve increases.

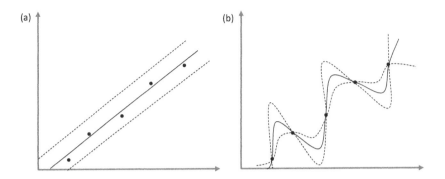

FIGURE 7.6   For results from (a) a physical experiment and (b) a deterministic computer experiment, approximate models of corresponding uncertainty.

Consequently, somewhere in the design space, a technique to minimize a deterministic computer code's worst-case estimation uncertainty is to strive to make points as spaced out in the design space as possible. This technique would minimize the distance between an observed result and any new position we choose to predict.

The variations between the actual model based on the physical experimental data and computer simulation models are shown in Figure 7.6. Consider a very simple scenario where we are calculating an output from a single input factor. Because the underlying assumptions about the data vary, different model choices and related uncertainty can be obtained. For the physical experiment data in Figure 7.6a, a linear regression line with predicted intercept and slope is the assumed model. The magnitude of each deviation is used to approximate the normal variance in the output, as the measurements do not lay exactly on the best-fitting line. If, on the other hand, as seen in Figure 7.6b, the data come from a deterministic simulation experiment, then an interpolator model is chosen. Such that the best approximate value for input is matched with a data point acquired from the machine code. As we have an exact value at that position for the computer code results, the variance for the simulator is zero, and the uncertainty bands around the function shrink to zero. Note that as we step away from any observation, the uncertainty about the predicted curve increases. Consequently, somewhere in the design space, a technique to minimize a deterministic computer code's worst-case estimation uncertainty is to strive to make points as spaced out in the design space as possible. This technique would minimize the distance between an observed result and any new position we choose to predict.

7.4 MODEL VERIFICATION AND VALIDATION

Models and simulations have been applied to practical system studies to solve problems and assist in decision-making requiring higher and higher precision and reliability standards. The creators and users of these models, decision-makers using data derived from the outcomes of these models, and the people impacted by decisions based on these models are all rightly concerned with whether a model and its outcomes are "correct." This concern is resolved by verification and validation of the model. Verification generally refers to the correct execution of the model for the intended purpose, whereas validation refers to the model's substantiation within the domain in question. The quality assurance of the models has become more and more relevant through mathematical verification and physical validation.

Many numerical modelers have followed a convenient validation method to satisfy the need for quality assurance, which is focused on comparing numerical model solutions to a limited number of field or laboratory evidence available at the time. The model is believed to have been validated as long as a fair or reasonable agreement is reached. Although the fair or reasonable agreement between numerical solutions and physical measurements in the lab or in the field can usually be accomplished by adjusting (or tuning) the model parameters, more and more scientists and model users have realized that such tuning is not validation but calibration of model parameters. The carefully "checked" and tuned models have often been found to have errors or imperfections, backed by the near-perfect fit between the simulations and the field observations. In the long process of mathematical derivations or manipulations, implementations of numerical solution schemes, applications of special features for speeding up/stabilizing, calculation algorithms, coding, mistakes, or errors may have occurred. It is difficult to find any of these faults or imperfections.

A mathematical method has been used to reduce the modelers' effort to fine-tune the model parameters, particularly if many model parameters have to be tuned one by one by trial and error. It is called "parametric identification." It is focused on the idea that the differences between calculated data and solutions to the model are reduced. The total or sum of the weighted average error between numerical model results and field data at all or selected data points can be minimized by applying an optimization analysis to obtain the parameters' values. Indeed, this technique will save time and effort when conducting parametric tuning. However, it is recommended for model developers and users to carefully test those tuned parameters' values to ensure that they are physically appropriate and then to use this effective parametric calibration method. There should be a thorough model validation followed.

It is essential that for each use or purpose of a model, a simulation model's verification and validation is carried out. If a simulation model aims to address a variety of questions about each question, the model's validity must be determined. The creators and users of simulation models, decision-makers who use the knowledge derived from these models, and the individuals influenced by the decisions based on those models are all correctly concerned with the question of models' "correctness" for each question dealt with by modeling. To define the domain of a model's intended applicability, numerous sets of experimental conditions are typically needed. A model may be valid for one set of experimental conditions and invalid for another. A set of experimental conditions includes a set of values for variables that describe the applicability domain. For a set of experimental conditions, a model is considered accurate if the model's accuracy is within its acceptable range of accuracy, which is the accuracy expected of the model for its intended purpose. Typically, this involves defining the output variables of interest of the model (i.e. the model variables used to address the questions that the model is being designed to answer) and then determining their appropriate range of accuracy. Before starting model development or early in the model development process, the appropriate range of accuracy of a model should be specified. If the interest variables are random variables, the random variable properties and functions, such as means and variances, are typical of primary interest and used to determine the model's validity. Before having a satisfactory valid model, many iterations of a model are typically created. The proof that a model is correct, i.e. carrying out model verification and validation, is usually considered a method and is usually part of the (total) phase of model creation.

7.4.1 Various Model Validation Techniques

Hypothesis testing and the validation of mathematical and computational models are fundamental to the science and engineering phase. There is no reason to assume that a model, which has been built to serve as a convenient representation of physical and/or human-made processes, has met its functional intent without a validation method. Thus, validation is the outcome of the optimistic justification of the portrayal of relevant characteristics in the real world by a model. Different users and developers have implemented various validation methods and tests for model verification and validation to their taste. These may be used subjectively or objectively. We mean using a mathematical procedure or statistical test, e.g. hypothesis tests or confidence intervals, when we refer to objectivity invalidation. Generally, a variety of strategies are used. In no order of choice, some of the main strategies today are introduced alphabetically in the following.

Comparison with other models: Different outcomes (e.g. outputs) of the validated simulation model are compared with other (valid) models. For instance, (1) simple simulation model instances are compared with known analytical model results, and (2) the simulation model is compared with other simulations.

Degenerate tests: The degeneracy of the model's behavior is evaluated by a good set of input values and internal parameters. For instance, when the arrival rate is greater than the service rate, does the average number in a single server's queue continue to increase over time?

Event validity: The "events" of simulation model occurrences are compared with those of the existing system to decide if they are identical. Compare the number of fires in a fire department simulation to the real number of fires, for instance.

Extreme condition tests: For every extreme and impossible combination of factors in the system, the model structure and outputs should be plausible. If in-process inventories, for example, are zero, production output should typically be zero.

Face validity: Individuals with knowledge of the method are asked if the model is fair and/or its actions are reasonably expected. For instance if the reasoning in the conceptual model is right and if the input–output relationship of the model is accurate.

Graphical visualization: The operational activities of the model are shown graphically as the model travels through time, e.g. during a simulation run, the movements of parts through a factory are seen graphically.

Historical approaches: Rationalism, empiricism, and constructive economics are the three historical methods of validation. Rationalism requires a straightforward statement of the assumptions underlying a model and that they are readily accepted. From these assumptions, logic deductions are used to construct the right (valid) model. Empiricism requires empirical confirmation of any statement and consequence.

Internal validity: To assess the amount of (internal) stochastic variability in the model, several replications (runs) of a stochastic model are made. A significant amount of uncertainty (lack of consistency) may cause the model's outcomes to be uncertain. If the problem organization is typical, the appropriateness of the policy or method being examined may be challenged.

Multistage validation: The three historical approaches of rationalism, empiricism, and optimistic economics are merged into a multistage validation process. This technique of validation consists of (1) establishing the assumptions of the model on theory, observations, and general knowledge, (2) validating, where possible, the model's assumptions by empirically checking them, and (3) comparing (testing) the model's input–output relationships with the entire system.

Operational graphics: Values of different performance metrics, such as the number of queues and the percentage of busy servers, are shown graphically as the model runs through time, i.e. as the simulation model runs through time, the complex actions of performance indicators are visually shown to ensure that they behave correctly.

Predictive validation: The model is used to predict the system's behavior (forecast). Then, comparisons are made between the system's behavior and the model's forecast to determine if they are the same. System data may come from an operating system or be collected by performing system experiments, such as field tests.

Sensitivity analysis: This approach involves adjusting the input values and the model's internal parameters to determine the effect on the model's action or performance. In the model, the same relationships can happen as in the existing system. Only qualitative output directions and quantitative input and output directions and (precise) output magnitudes can use this method. Before using the model, certain parameters that are sensitive, i.e. cause major changes in the model's behavior or performance, should be made sufficiently accurate. (In model creation, this can require iterations.)

Traces: The actions of various types of entities in the model are tracked (followed) through the model to decide if the model's logic is accurate and whether the requisite precision is obtained.

Turing tests: Whether they can differentiate between system and model outputs, individuals informed about the system's operations being modeled are asked.

Validation of historical data: If historical data exist (e.g. data obtained on a system explicitly for the creation and testing of a model), part of the data are used to create the model, and the remaining data are used to assess (test) whether the model acts as the system.

For testing simulation models, there are two basic approaches: static testing and dynamic testing. The computer program is evaluated in static testing to determine if it is accurate using such techniques as organized walkthroughs, proof of validity, and analyzing the model's structural properties. The computer model is run under various conditions in dynamic testing. The values obtained (including those created during execution) are used to decide whether the computer model and its implementations are correct.

Operational validation determines whether the simulation model's output behavior has the precision needed for the intended function of the model over the domain of the intended applicability of the model. This validation is where much of the research and assessment for validation takes place. Since the simulation model is used in operational validation, any flaws found could be triggered by what has been produced in any of the steps involved in creating the simulation model, including creating the theories of the system or the possession of invalid data.

Comparisons of the model's and system's output behaviors for many different sets of experimental conditions are typically needed to achieve a high degree of trust in a simulation model and its performance. Thus, if a system is not measurable, as is always the case, a high degree of trust in the model can typically not be achieved. In this case, model performance behavior(s) should be investigated as extensively as possible, and where possible, comparisons should be made with other relevant models. In comparing the simulation model output behavior to either the system output behavior or other model output behavior, there are three fundamental methods used: (1) the use of graphs to make a subjective judgment, (2) the use of confidence intervals to make an objective decision, and (3) the use of hypothesis tests to make an objective decision. The use of confidence intervals or hypothesis tests for comparisons is preferable because they allow rational decisions.

7.4.2 Automation Tools for Model Verification and Validation

Validation of models is an iterative method that continues for at least as long as the structure described by the model is under development. However, if model validation involves a considerable manual effort, validation activities will be rare during system creation. Organizations fail to incorporate, validate, and maintain a vast amount of test findings from a range of different sources. The amount of time taken to conduct the product verification and validation process increases, as the amount of test procedure results produced from different sources and manual simulators increases. Consequently, there are administrative costs and delays in reporting problems that further complicate the product growth phase.

Therefore, straightforward and highly automated methods for evaluating a model's validity regarding the current system configuration are critical challenges in developing methods for validating the model effectively. A methodology for continuous model verification and validation is important to justify model-based design decisions during the entire system development process. When new data become available, model verification and validation operations are essentially quick and simple to reiterate. The use of suitable approaches, automation design, and related tools has tremendous potential for verification and validation process optimization.

The main objective is the validation of multivariate computational models that represent unknown conditions and/or results. The instability in a model or data set may derive from stochasticity, model parameter and input data uncertainty, calculation uncertainty, or other potential sources of uncertainty. Any of the following data simulation systems may provide quantifiable levels of variance that we would like to test based on a collection of validation data: neural networks and AI models, ML models, Gaussian-phase regression models, polynomial chaos and other surrogate models, spatial and time series stochastic models, physics-dependent models (usual solutions to differential equations), engineering-based models (which are sufficiently abstracted mechanics-based models), Monte Carlo simulation models, and more. Model performance uncertainties may be quantified by uncertainty propagation techniques (that may include verification, calibration, and validation).

The degree to which the model represents the physical system is quantified through relevant verification and validation measures concerning the model's intended use. Hence, a concrete formulation of the expected usage is required for both model development and verification and validation process. For example, automating historical data verification and validation, i.e. verification and validation against existing dimensions, comprises the automation of several necessary steps in the verification and validation phase, starting from the specification of the model's operation realm visualization and analysis of verification and validation effects. Following moves are considered necessary for efficient historical data model validation, and all need to be automated to optimize automation in model verification and validation:

a. The model input dependencies need to be defined. The model's realm of operability is a space spanned by the feasible values of input variables. A consistent description of this space is important for coverage and device-level validity metrics.

b. Description of validation amounts. The machine variables that collectively characterize the model's validity need to be established before any validation simulation can be performed.

c. Recognition of steady-state activity. The steady-state activity must be defined in the calculations for steady-state validation operations. The detection technique can be performed before or during simulation if an entire flown mission is simulated. Furthermore, identifying steady-state activity before simulation offers the potential to use coverage criteria to prioritize which simulations are most suited for validation.

d. Computation of steady-state validity parameters. The obvious object of verification and validation activities is to determine the model's validity within its defined operation area: the model's domain of validity. The chosen validation metrics are the basis for formulating this domain.

e. Visualization of validation outcomes. Once developed, a model's validity needs to be conveyed to the model's consumer comprehensibly. This feedback is a daunting job in the case of dynamic models of high-dimensional domains of operations.

While ML models with the significant predictive ability and actionable insights are reliable, the increased sophistication of these models poses specific validation process challenges. Besides, some ML models are configured to be redeveloped on a dynamic basis automatically as new data become available, making the validation process much more complicated. The ML infrastructure needs to be able to store the model modifications separately along with the input/output data before validation can be completed, or the best choice is to simplify the validation process and the automatic reconstruction process. For this to work, customized automatic validation processes need to be developed and integrated with the ML and digital twin models to be effectively deployed in the industry.

References

Andrén, F. P., T. I. Strasser, J. Resch, et al., 2019, Towards automated engineering and validation of cyber-physical energy systems, *Energy Informatics*, Vol. 2, No. Suppl 1, p. 7. https://doi.org/s42162-019-0095-x.

Aruldhas, G., 2008, *Classical Mechanics*, Asoke K. Ghosh, PHI Learning Private Limited, New Delhi.

Ash, R. B., 2008, *Basic Probability Theory*, Dover Publications, Inc, New York.

Bécue A., E. Maia, L. Feeken, P. Borchers, and I. Praça, 2020, A New Concept of Digital Twin Supporting Optimization and Resilience of Factories of the Future, *Applied Sciences*, Vol. 10, p. 4482. doi:10.3390/app10134482.

Beisbart, C., and S. Hartmann, 2011, *Probabilities in Physics*, Oxford University Press, Oxford.

Belsley, D. A., E. Kuh, and R. E. Welsch, 2004, *Regression Diagnostics Identifying Influential Data and Sources of Collinearity*, John Wiley & Sons, Inc., Hoboken, NJ.

Ben-Menahem, Y., l. M. Hemmo, 2012, *Probability in Physics*, Springer-Verlag, Berlin and Heidelberg.

Bes, D. R., 2012, *Quantum Mechanics: A Modern and Concise Introductory Course*, Springer-Verlag Berlin, Heidelberg.

Bolkvadze, G. R., 2003, The Hammerstein-Wiener Model for Identification of Stochastic Systems, *Automation and Remote Control*, Vol. 64, No. 9, pp. 1418–1431.

Brandimarte, P., 2014, *Handbook in Monte Carlo Simulation Applications in Financial Engineering, Risk Management, and Economics*, John Wiley & Sons, Inc., Hoboken, NJ.

D'Auria, R., and M. Trigiante, 2012, *From Special Relativity to Feynman Diagrams A Course of Theoretical Particle Physics for Beginners*, Springer-Verlag, Italy.

Daniel R. B., 2012, *Quantum Mechanics A Modern and Concise Introductory Course*, Springer-Verlag, Berlin, Heidelberg.

Danielsen-Haces, A., 2018, Digital Twin Development: Condition Monitoring and Simulation Comparison for the ReVolt Autonomous Model Ship, Master's Thesis, Norwegian University of Science and Technology, Trondheim, Norway.

Das, A., 2008, *Lectures on Quantum Field Theory*, World Scientific Publishing Co. Pte. Ltd, New Jersey.

Datta, S. P. A., 2017, Emergence of Digital Twins, *Journal of Innovation Management*, Vol. 5, pp. 14–34.

Dawei, J., 2011, The Application of Data Mining in Knowledge Management, 2011 International Conference on Management of e-Commerce and e-Government, IEEE Computer Society, pp. 7–9. doi:10.1109/ICMeCG.2011.58

Deriglazov, A., 2010, *Classical Mechanics: Hamiltonian and Lagrangian Formalism*, Springer-Verlag, Berlin, Heidelberg.

Dodiya, D., 2014, An Analysis of Use of Automated Tools for Improving the Process of Software Verification and Validation in a Midwestern Company, Dissertations and Theses @ UNI. 30. https://scholarworks.uni.edu/etd/30.

Dütsch, M., 2019, *From Classical Field Theory to Perturbative Quantum Field Theory*, Springer Nature, Switzerland.

Fayyad, U., G. Piatetsky-Shapiro, and P. Smyth, 1996, From Data Mining to Knowledge Discovery in Databases, *AI Magazine*, Vol. 17, No. 3, pp. 37–54.

French, A. P., 1971, *Newtonian Mechanics*, William Clowes & Sons Ltd, Beccles, Colchester and London.

Galvanin F., M. Barolo, and F. Bezzo, 2010, A Framework for Model-based Design of Experiments in the Presence of Continuous Measurement Systems, 9th International Symposium on Dynamics and Control of Process Systems (DYCOPS 2010) Leuven, Belgium, July 5–7, 2010, pp. 571–576.

Garlapati, V. K., and L. Roy, 2017, Utilization of Response Surface Methodology for Modeling and Optimization of Tablet Compression Process, *Journal of Young Pharmacists*, Vol. 9, No. 3, pp. 417–421.

Ghosh, A., 2018, *Conceptual Evolution of Newtonian and Relativistic Mechanics*, Springer Nature Singapore Pte Ltd, Singapore.

Glynn, P. W., and Y. L. Jan, 2013, *Stochastic Simulation and Monte Carlo Methods: Mathematical Foundations of Stochastic Simulation*, Springer-Verlag Berlin, Heidelberg.

Han, J., M. Kamber, and J. Pei, 2012, *Data Mining Concepts and Techniques*, Third Edition, Elsevier Inc.

Heisenberg, w., 1927, Originally published under the title, "Uber den anschaulichen Inhalt der quantentheoretischen Kinematik und Mechanik," *Zeitschrift for Physik*, Vol. 43, pp. 172–198 (1927); reprinted in Dokumente der Naturwissenschaft, 4, 9–35 (1963); translation into English by J.A.W. and W.H.Z., 1981.

Huang, K., 1987, *Statistical Mechanics*, John Wiley and Sons Inc, Toronto.

Jaynes, E. T., 2003, *Probability Theory the Logic of Science*, Cambridge: Cambridge University Press, New York.

Landau L. D. and E. M. Lifshitz, 1994, *The Classical Theory of Fields Translated from the Russian by Morton Hamermesh*, Pergamon Press pic.

Landau, R. H., M. J. Páez, and C. C. Bordeianu, 2007, *Computational Physics Problem Solving with Computers*, 2nd, Revised and Enlarged Edition, WILEY-VCH Verlag GmbH & Co KGaA, Weinheim.

Lavis D.D, 2011, An Objectivist Account of Probabilities in Statistical Mechanics, in *Probabilities in Physics*, Beisbart C. and S. Hartmann (eds), Oxford University Press, Oxford, pp. 51–81.

Libof R. L., 1980, *Introductory Quantum Mechanics*, Addison-Wesley Publishing Company Inc, New York.

Lönn, D., 2008, Robust Design – Accounting for Uncertainties in Engineering, Thesis No. 1389, Linköping University, Linköping, Sweden.

Lùcio, L., B. Barroca, and V. Amaral, 2010, *A Technique for Automatic Validation of Model Transformations*, D. C. Petriu, N. Rouquette, Ø. Haugen (Eds.), MODELS 2010, Part I, Springer-Verlag, Berlin, Heidelberg, pp. 136–150.

MacKay, D. J. C., 2005, *Information Theory, Inference, and Learning Algorithms*, Cambridge University Press.

Mahadevan, L., and E. H. Yong, Probability, Physics and the Coin Toss, *Physics Today*, July 2011, pp. 66–67.

Maña, C., 2017, *Probability and Statistics for Particle Physics*, Springer International Publishing AG, Cham, Switzerland.

McMahon, D., 2008, *Quantum Field Theory Demystified*, The McGraw-Hill Companies, Inc.

Michelsen, Eric L., 2019, Funky Statistical Mechanics Concepts, www.elmichelsen.physics.ucsd.edu.

Morin, D., 2008, *Introduction to Classical Mechanics with Problems and Solutions*, Cambridge University Press, New York.

Mun, J., 2006, *Modeling Risk: Applying Monte Carlo Simulation, Real Options Analysis, Forecasting, and Optimization Techniques*, John Wiley & Sons, Inc., Hoboken, NJ.

Myers, R. H., D. C. Montgomery, and C. M. Anderson-Cook, 2016, *Response Surface Methodology: Process and Product Optimization Using Designed Experiments*, John Wiley & Sons, Inc., Hoboken, NJ.

Øvern, A., 2018, Industry 4.0- Digital Twins and OPC UA, Master's Thesis, Norwegian University of Science and Technology, Trondheim, Norway.

Pan, Y. and S. A. Billings, 2006, Model Validation of Spatiotemporal Systems using Correlation Function Tests. Research Report. ACSE Research Report No. 915. Automatic Control and Systems Engineering, University of Sheffield.

Park, G-J., T-H. Lee, K. H. Lee, and K-H Hwang., 2006, Robust Design: An Overview, *AIAA Journal*, Vol. 44, No. 1, pp. 181–191.

Rasheed A., O. San, and T. Kvamsdal, 2020, Digital Twin: Values, Challenges and Enablers from a Modeling Perspective, *IEEE Access*. doi:10.1109/ACCESS.2020.2970143.

Rivas, M., 2002, *Kinematical Theory of Spinning Particles: Classical and Quantum Mechanical Formalism of Elementary Particles*, Kluwer Academic Publishers, New York.

Robert Hällqvist, R., M. Eek, R. Braun, and P. Krus, 2016, Methods for Automating Model Validation: Steady-State Identification Applied on Gripen Fighter Environmental Control System Measurements, 30th Congress of the International Council for the Aeronautic Sciences (ICAS).

Roos, D., U. Adam, and C. Bucher, 2006, Weimar Optimization and Stochastic Days 3.0, November 23–24, 2006.

Roy, R. K., 2001, *Design of Experiments Using Taguchi Approach*, John Wiley, and Sons Inc., New York.

Rud, O. P., 2001, *Data Mining Cookbook: Modeling Data for Marketing, Risk, and Customer Relationship Management*, Published by John Wiley & Sons, Inc, New York.

Wang, S. S. Y., 2009, Verification and Validation of Free Surface Flow Models 1, in *Verification and Validation of 3D Free-Surface Flow Models*, S. S. Y. Wang, P. J. Roche, R. A. Schmalz, Y. Jia, and P. E. Smith (Eds.), American Society of Civil Engineers, Reston, Virginia.

Sanchez, S. M., 2000, Robust Design: Seeking the Best of all Possible Worlds, Proceedings of the 2000 Winter Simulation Conference.

Sargent, R. G., 2011, Verification and Validation of Simulation Models, Proceedings of the IEEE 2011 Winter Simulation Conference.

Sargent, R. G., 2015, An Introductory Tutorial on Verification and Validation of Simulation Models, Proceedings of the IEEE 2015 Winter Simulation Conference.

Scheck, F., 2012, *Classical Field Theory: On Electrodynamics, Non-Abelian Gauge Theories and Gravitation*, Springer-Verlag Berlin, Heidelberg.

Schwinger, J., 2000, *Quantum Kinematics and Dynamics*, Westview Press, New York

Taeyoung L., M. Leok, and N. H. McClamroch, 2018, *Global Formulations of Lagrangian and Hamiltonian Dynamics on Manifolds: A Geometric Approach to Modeling and Analysis*, Springer International Publishing AG.

The thing about data. *Nature Physics*, Vol. 13, No. 717 (2017). https://doi.org/10.1038/nphys4238

Theodoridis, S., 2015, *Machine Learning: A Bayesian and Optimization Perspective*, Elsevier Ltd.

Torre, C. G., 2019, Introduction to Classical Field Theory. *All Complete Monographs*, 3, https://digitalcommons.usu.edu/lib_mono/3.

Wang, Z., 2020, Digital Twin Technology, in Industry 4.0–Impact on Intelligent Logistics and Manufacturing, http://dx.doi.org/10.5772/intechopen.80974

Widom, B., 2002, *Statistical Mechanics: A Concise Introduction for Chemists*, Cambridge University Press, New York.

Yang, H., and E. K. Lee, 2016, *Healthcare Analytics: From Data to Knowledge to Healthcare Improvement*, John Wiley & Sons, Inc., Hoboken, NJ.

Zang, C., M. I. Friswell, and J. E. Mottershead, 2005, A Review of Robust Optimal Design and its Application in Dynamics, *Computers and Structures*, Vol. 83, pp. 315–326.

Zee, A., 2010, *Quantum Field Theory in a Nutshell*, Second edition, Princeton University Press, Princeton, NJ.

Index

Printed in the United States
by Baker & Taylor Publisher Services